Modeste Defeu
Félix Meutchieye
Yacouba Manjeli

Biodiversity of Donkeys in the Highlands of Northwest Cameroon

Modeste Defeu
Félix Meutchieye
Yacouba Manjeli

Biodiversity of Donkeys in the Highlands of Northwest Cameroon

ScienciaScripts

Imprint
Any brand names and product names mentioned in this book are subject to trademark, brand or patent protection and are trademarks or registered trademarks of their respective holders. The use of brand names, product names, common names, trade names, product descriptions etc. even without a particular marking in this work is in no way to be construed to mean that such names may be regarded as unrestricted in respect of trademark and brand protection legislation and could thus be used by anyone.

Cover image: www.ingimage.com

This book is a translation from the original published under ISBN 978-3-639-60825-0.

Publisher:
Sciencia Scripts
is a trademark of
Dodo Books Indian Ocean Ltd. and OmniScriptum S.R.L publishing group

120 High Road, East Finchley, London, N2 9ED, United Kingdom
Str. Armeneasca 28/1, office 1, Chisinau MD-2012, Republic of Moldova, Europe
Managing Directors: Ieva Konstantinova, Victoria Ursu
info@omniscriptum.com

Printed at: see last page
ISBN: 978-620-9-73986-6

TABLE OF CONTENTS

DEDICATION

To

J My father NZUDIE JEAN-CLAUDE

J My mother, CHEKAM MICHELINE

J My older brother TALLE PASCAL HERVE

ACKNOWLEDGMENTS

This work is the result of the efforts, care, and encouragement that have accompanied me from its conception and implementation through to its completion. For this reason, I cannot fail to express my deepest gratitude:

- To Almighty God for His wonders and His mercy;
- To my thesis advisor, Prof. MANJELI Yacouba, for agreeing to supervise this work with great patience, despite his many responsibilities. Thank you for the extensive resources and your continued trust;
- To Mr. MEUTCHIEYE Félix, for his advice, encouragement, and especially for his patience in supervising this work;
- To the entire faculty and staff of the University of Dschang, and in particular, the Department of Animal Production, who generously shared their knowledge to support my education. These include Professors Joseph TCHOUMBOUE, Alexis TEGUIA, André ZOLI, PAMO TEDONKENG, Florence FONTEH, Jules Roger KUIATE, François KAMAJOU, Ferdinand NGOULA, and Henri DEFANG; and Doctors Raphael KANA, Fernand TEDONKENG, and Augustave KENFACK;
- To my dear parents, Jean-Claude NZUDIE and Micheline CHEKAM, and my older brother Pascal TALLE, without whom I would never have made it this far. I thank them for the tremendous moral and material support they have given me throughout my studies, from my earliest years until today, and for respecting my choice;
- To my grandmother, Mama SIMO Julienne; my uncles, Mr. FEUHOUO François and his wife, Mrs. GUEMCHUEN Edwige; Mr. TINDJOU Christophe; Mr. FOKAM Bienvenu; Mr. BOUOPDA Théodore; and my aunts: Mrs. TALLA Suzanne, Mrs. TALLA

Thérèse, Mrs. NGUEMWO Hélène, WOKAM Léopoldine and her husband KUATCHE Severin, DONGNE Flore, SOFEU Justine, for their constant prayers and moral support;

- To my brothers and sisters: Pascal Talle, Verlaine Woguem, Stéphanie Kenmogne, Edith Tchomtseu, Arnold Mabou, and Ariane Djemmo—may this work serve as a source of inspiration for them throughout their academic journey;
- To my neighbors in the housing project and the Saint Vincent de Paul Choir, for their prayers;

To the staff of the BUI delegation in the Northwest. To CHING Loretta, my field guide, as well as the farmers, I offer my deepest gratitude;

To my classmates in the doctoral program in Biotechnology and Animal Production at the , particularly those in my specialization: I am thinking of DJOMTCHAIGUE Herbert, DJOUFACK Yannick, and TEBOURDANNE Gaston; may they find here the expression of my deep gratitude;

- To my senior classmates, NKENFACK Patrick, AYAGUIRWE Rodrigue, TANGOMO Armel, and DALOUM Solidaire, for their advice and for reviewing this document;
- To my friends JOUONZO Landry Gautier, LONGANG Saubaber, and BOULOU MENGUE Raphael;
- Finally, I would like to express my special thanks to all those who have helped me in any way, encouraged me, and supported me throughout all the years of my studies.

ABSTRACT

The study was conducted from September to December 2014 in the highlands of northwestern Cameroon. The overall objective was to contribute to a better understanding of donkey diversity with a view to its conservation and genetic improvement, and more specifically: to evaluate morphological characteristics; to evaluate body measurements of donkeys; to establish, based on body measurements, equations for predicting the live weight of donkeys; to estimate biometric indices of donkeys based on body measurements and to analyze genetic variability, and to establish the genetic structure and relationships among donkey populations. To this end, a sample of 174 adult donkeys (96 males and 78 females) was observed in 7 localities and 3 markets within the study area. The results show that there is high phenotypic variability among donkeys in the highlands of northwestern Cameroon. In terms of coat color, gray is dominant (86.2%) and burnt bay is less common (0.5%). Body measurements yielded the following values: height at the withers 99.73±3.57 cm, chest circumference 108.66±5.08 cm, chest width 25.85±2.62 cm, chest depth 48.80±3.97 cm, body length 110.07 ± 6.06 cm, height at the croup 103.28 ± 3.91 cm, cannon circumference 14.95 ± 1.85 cm, and live weight 128.68 ± 17.07 kg. The correlations are significant ($P<0.01$), positive, and nearly perfect between live weight and chest circumference ($r=0.99$). The live weight prediction equation LW = 2.34BC - 233.89 appears to be the best due to its simplicity and high coefficient of determination (R^2= 0.99). The profile and relative body indices are 0.91 ± 0.04 and 1.01 ± 0.05, respectively, and allow the donkeys of the highlands of northwestern Cameroon to be classified as long-bodied. The dactylothoracic index is 0.14 ± 0.04 and suggests that the donkeys are of the hypermetric type. The genetic diversity of the donkey population, as determined by principal component analysis (PCA), revealed that live weight and body length account for 59.76% of the genetic variability within the population. Discriminant factor analysis (DFA) shows that the population consists of three genetic types: genetic type C with high characteristics: (Height at the withers = 103.61 cm, Chest circumference = 118.61 cm, Live weight = 164.05 kg); genetic type A with average characteristics (withers height = 100.79 cm, chest circumference = 110.42 cm, live weight = 134.02 kg); and genetic type B with low characteristics (withers height = 97.00 cm, chest circumference = 103.30 cm, live weight = 111.13 kg). The phylogenetic tree clarified the relationships among the three genetic types: genetic types A and C form a single subgroup 1 that is distinct from subgroup 2 (genetic type B). The observed variability suggests that donkeys from the highlands of northwestern Cameroon can be improved through selection and crossbreeding techniques.

Keywords: Donkey, Biodiversity, measurements, biometric index, regression, Northwestern Highlands of Cameroon.

INTRODUCTION

The donkey (*Equus asinus*) is a domestic animal belonging to the Equidae family, which includes the horse and the zebra (Wilson, 1990). The donkey was domesticated thousands of years ago (Jussiau *et al.,* 1999; Jean-Pierre, 2009) and has contributed to the development of various civilizations.

In every country, it performs the same functions as the horse at a lower cost. The poorest farmers prefer it to the horse because it requires little food, adapts better to harsh environmental conditions than the horse, and is reputed to withstand tsetse-fly-infested regions better than zebu (Luurt, 2001). It is the most important domestic animal in several countries in North Africa and the Middle East (Smith and Pearson, 2005), because it can survive, reproduce, and produce meat and milk under harsh environmental conditions. Donkeys are overwhelmingly used for work; more than 95% of them are reportedly raised for this purpose (Chappez, 2000). They are also used for riding, as pack animals, for transport, and as draft animals (Tiringo, 2010);

Donkey meat can be considered a product of high nutritional value, characterized by a high content of unsaturated fatty acids—specifically polyunsaturated fatty acids—and contains a high level of essential amino acids, making it a healthy alternative to red meat (Polidori *et al.,* 2009); Some of these amino acids are precursors of antithrombotic factors (Kinsella, 1988).

Donkey milk has long been praised for its nutritional quality and its similarity to human milk (Carroccio *et al.,* 2000; Fanica, 2008; Guo *et al.,* 2007); it is used as a substitute for cow's milk for children who are intolerant to cow's milk proteins and as a natural food source in situations where breastfeeding is not possible (Tesse *et al.,* 2009). Various studies confirm that donkey milk is a natural product with good dietary and cosmetic qualities (Carroccio *et al.,* 2000; Piccione *et al.,* 2008; Vincenzetti *et al.,* 2008; Siméon, 2008), with good palatability and optimized intestinal calcium absorption (Schaafsma, 2003).

The global donkey population is estimated at over 43 million in 2005 (Bougler, 2005) and 41 million in 2006 (Kugler *et al.,* 2008), with over 9 million in Africa. China alone has over 10 million, and Ethiopia has over 5 million. Over the centuries, human needs have shaped these breeds, favoring certain morphological traits to produce uniform animals. Current morphological standards are the result of this (Siméon, 2008). There are 185 recognized donkey breeds worldwide (Kugler *et al.,* 2008).

In Cameroon, the donkey population is estimated at over 50,000 (INS, 2013), and donkey husbandry is practiced in almost all regions. Some studies have been conducted on the socio- al aspects of donkey husbandry (Ebangi and Vall, 1998; Vall *et al.,* 1996; Vall et al., 1998; Vall *et al.,* 2001). However, studies on the morphobiometric characteristics of donkeys in the highlands of Northwest

Cameroon are scarce. Yet, according to Bouchel *et al.* (1997), knowledge of morphobiometric characteristics is recognized as one of the strategies for analyzing domestic populations.

The overall objective of this study is to contribute to the study of donkey biodiversity in the highlands of Northwest Cameroon with a view to its genetic improvement and conservation. More specifically, the aim is to analyze:

- the morphological characteristics of the donkey species in the highlands of northwestern Cameroon;
- the body measurements of donkeys;
- the equations for predicting donkey live weight based on body measurements;
- biometric indices of donkeys based on body measurements;
- the genetic variability, structure, and genetic relationships among donkey populations in the highlands of northwestern Cameroon.

Following the literature review, the materials and methods used will be presented, followed by the main results and, finally, the main conclusions and future directions.

CHAPTER I: LITERATURE REVIEW

1.1 Definition of Concepts

> Systematics

Systematics is the discipline that assigns a specific place to a living organism within a classification system composed of nested categories (Pellegrini, 1999). These categories, in descending order of magnitude, are: Kingdom, Phylum, Class, Order, Family, Genus, and Species. This terminology is attributed to the Swedish naturalist Linnaeus (1707–1778), the first to propose a classification of plants and animals based on their morphological types.

Living organisms are thus expressed through a series of levels, the last two of which—Genus and Species—serve to universally identify them.

> Phylogeny

In 1866, the German biologist Haeckel coined the term "phylogeny" to describe the lineage that unites living beings.

Phylogeny is the study of kinship relationships among different living beings in order to understand the evolution of living organisms (Darwin, 1859).

> Population.

A population is a group of individuals of the same species living in the same territory and reproducing exclusively among themselves (De Rochambeau, 1990).

According to Aubineau *et al.* (2002) and De rochambeau (1990), a population is a group of individuals of the same species living in the same territory and capable of interbreeding. The traditional or primary population is defined as a group of domestic animals directly descended from a wild species and characterized by a high degree of phenotypic variability (Machado *et al.*, 1992).

> **Domestication:** Helmer (1979) offers the following definition: domestication is the control of an animal population through the isolation of the herd, resulting in the loss of panmixia, the suppression of natural selection, and the application of artificial selection based on specific traits, whether behavioral or structural. The living animals effectively become the property of the human group and are entirely dependent on humans.

According to Danchin (2002), the primary effect of domestication is the emergence of variations in physical appearance that do not exist in the wild (color genes, hornedness, etc.)

> **Species**

According to Mahaman (1986), a species is a natural group of individuals that share fairly similar morphological and physiological traits and are capable of mating to produce offspring that are indefinitely fertile with one another and with their parents.

> **Breed**

A breed can be defined as: a set of common traits that are heritable from one generation to the next (Habault, 1974).

The term "breed" refers to the group of descendants of a family or a natural group of individuals sharing a set of common physical characteristics (Stansfield, 1975).

A breed consists of a group of similar individuals belonging to the same species and sharing a certain number of common hereditary traits (Audiot, 1995).

Pellegrini (1999) noted that the concept of a breed would then correspond to "a local or regional population adapted to a specific production system and exhibiting a certain number of common characteristics.

> **Genetic variability**.

According to Verrier *et al.* (2005), genetic variability can be defined, at a given locus, as the diversity of alleles found and, across a set of loci, as the diversity of alleles and their combinations." Phenotypic variability is the expression of genetic variability (Lauvie, 2007).

> **Genetic diversity.**

Genetic diversity is the total genetic information contained in all living organisms and corresponds to the variability of genes and genotypes between species and within each species (Fadlaoui, 2006). For livestock, diversity is measured by the genetic variability between breeds of the same species and within the breeds of that species (Fadlaoui, 2006). It naturally gives rise to a very wide variation in the phenotypes of these animals (Tavernier and Lizeau, 2000).

> **Phenotypic characterization.**

According to Bogart (1965), the phenotype is the description of an animal: external appearance, performance measured through testing, and resistance to disruptive factors, without knowing whether the variation of the trait from the mean is due to the environment or heredity. There are two categories of phenotypic characteristics:

> Qualitative traits, whose variation is discrete (e.g., pigmentation of the , coat color, presence or absence of horns, etc.).

> Quantitative or biometric traits, whose variation is continuous (e.g., animal weight) or discrete, taking on a numerical value in the animals that express them, are due to many genes, each producing a small effect.

The principle of characterizing an animal population is based on examining the morphological profile of adult animals, focusing on two aspects: phenotypic profiling, such as wool color, and biometric profiling, such as body length and height (Boumaza, 1974; Flamant, 1988).

The study of the morphological characteristics of animal populations aims to identify breeds and distinguish between them (Ouragh *et al.,* 2002), as well as to determine the degree of genetic variability for the improvement and maintenance of genetic diversity, which enables breeders to select animals or create new breeds to cope with environmental changes (Traore *et al.,* 2006; De Rechambeau *et al.,* 2003)

1.2 General Information on Donkeys

1.2.1 Origin and domestication of the donkey

1.2.1.1 Origin of the modern donkey

The first equids appeared during the Pleistocene (about one million years ago) (Siméon, 2008). Morphological data, combined with cytogenetic evidence, confirm that there are currently three main lineages represented by horses, donkeys, and zebras, which successively evolved from the Quaternary *Equus* (Siméon, 2008). The African wild ass (*Equus africanus*) is the ancestor of the domestic donkey (*Equus asinus*) (Linnaeus, 1758). There are three subspecies (Groves 1966, 1986; Haltenorth and Diller, 1980): the North African wild ass (*Equus asinus atlanticus*), the Somali wild ass (*E. africanus somaliensis*), and the Nubian wild ass (*Equus africanus africanus*).

1.2.1.2 Domestication of the donkey

Several studies and research projects have been conducted to determine the exact time and location of the domestication of the donkey. It is generally accepted that the domestication of the donkey dates back to 5000 BCE in the Nile Valley (Jussiau *et al.,* 1999; Jean-Pierre, 2009). In 2004, researchers from the CNRS's Laboratory of Alpine Ecology demonstrated through genetic studies that the common donkey (*Equus asinus*) was likely first domesticated in Northeast Africa. The ancestors of the common donkey are thus believed to be the Nubian wild ass (*Equus africanus africanus*) and the Somali wild ass (*Equus africanus somaliensis*) (Albano *et al.*, 2004). In 2007, donkey remains

found in an Egyptian tomb at Abydos showed characteristic skeletal deformities resulting from their use as pack animals, which do not exist in wild donkeys (Rossel *et al.*, 2008). The donkey was the first equid to be domesticated (Jean Pierre, 2009); but this domestication occurred well after that of cattle, sheep, and goats, which date back to the seventh and eighth millennia BCE. According to the findings of Albano *et al.* (2004), today's donkeys are not descended from the Asian lineage, but from the African lineage. Detailed analysis of mtDNA reveals that the Nubian donkey and the Somali donkey were domesticated separately. Researchers suggest that this domestication was a response to new transportation needs among pastoral societies in Northeast Africa, which were facing the desertification of the Sahara (between 5,000 and 7,000 years ago).

1.2.2 Classification of the domestic donkey

The scientific name for the donkey (*Equus asinus*) actually refers to the African wild ass, which belongs to the same genus as the horse, a genus that includes six species:

- The horse (*Equus caballus)*
- The donkey (*Equus asinus)*
- The onager (*Equus hemionus)*
- The quagga (*Equus quagga*)
- The mountain zebra (*Equus zebra)*
- Grevy's zebra (*Equus grevyi)*

The classification of the domestic donkey is as follows:

Kingdom: **Animal**

Phylum: **Chordata**

Class: **Mammals**

Order**: Perissodactyla**

Family: **Equidae**

Genus: **Equus**

Species: **Equus asinus**

Source: Yarkin, 1962

1.3 Significance and importance of donkey breeding

1.3.1 Significance of donkey husbandry

1.3.1.1 Sociocultural significance

The donkey has strong symbolic connotations. The value of the donkey varied across societies. Indeed, it is a symbol of poverty and humility in the Christian religion, and in India, Hindu celestial deities are carried on its back. It is also a symbol of stupidity and an object of ridicule: due to its long association with the humblest members of rural societies, its appearance—which is seen as slow and unalert—and its stubbornness. The Romans attributed medicinal properties to donkey blood, sweat, and urine. The donkey also holds particular significance for certain traditional religions. As early as the Old Testament, the sacrifice of a donkey reminded the Jews of their exodus from Egypt. In Senegal, on a mystical level, the donkey's head is said to be used to exorcise the demons of hunger (Tiringo, 2010).

1.3.1.2 Economic importance

The domestic donkey has long served the same purposes as the horse, but it was less expensive because it required very little food; poorer farmers therefore preferred it to the horse, hence its nickname, "the poor man's horse" (Farissier, 2007). Today, it is still used for transportation and agriculture, particularly in Africa, but it is also finding new roles in the most industrialized countries, where, no longer recognized as a working animal, it is increasingly taking on the role of a recreational companion (Farissier, 2007).

- **Transportation of people and goods:** Donkeys are sometimes used to transport people (Chappez, 2000). They are used to transport goods (sand, firewood, crops), carrying loads of up to 80 to 100 kg; transporting people, or pulling donkey-drawn carts. These journeys can sometimes cover long distances (15–20 km) (Roamba, 2014).

- **Man's working companion:** Since ancient times, the donkey has been man's working companion, pulling the plow to till the soil before sowing, threshing wheat or barley after the harvest, and working all day long on the threshing floor to separate the grains from the stalks (Chappez, 2000). Of all animals used for draft work, the donkey exerts the greatest pulling force relative to its weight, 10 to 16% of its body weight.A donkey with a live weight between 100 and 150 kg provides, on average, the same effort as an ox with a live weight between 325 and 450 kg (Vall, 1998).

- **Donkey races**: The first donkey races appear to have been organized in the last century in the Vendée region. The donkeys participating in the races were either ridden directly or harnessed.

- **The donkey, a circus animal:** In the past, there were more circus acts performed with donkeys than with horses. As companions to the clowns, donkeys were there to make the audience laugh (Chappez, 2000).

- **The donkey as a pet:** Today, in industrialized countries, the donkey has become a pet just like the dog and the cat.

- **Therapeutic Activities:** Donkeys can also be used as therapy animals (Farissier, 2007). Donkey therapy is thus similar to equine therapy, with which it shares many commonalities simply because both involve working with equines. However, donkeys possess psychological and physical characteristics that distinguish them from horses in a therapeutic setting. They are highly sensitive animals with a strong need for contact, seeking in humans not so much a guide as a companion. More cautious, it needs to build a relationship before granting its trust. Its morphology also distinguishes it from the horse in the therapeutic approach. Its role is that of a mediator between the person with a disability and the care staff (Siméon, 2008).

Donkey breeding can be a profitable business in the Sahel regions. Below certain isohyets, donkey breeding becomes a growing problem due to humidity-related infections (RIM, 1992). It is therefore more practical for donkey users to purchase animals from the north and replace them at the end of their working lives. Sahelian countries such as Niger and Mali have a significant trade in donkeys, especially males, to communities located to the south (Blench, 2012).

The extent to which donkeys are consumed is likely underestimated, as this is sometimes considered taboo in certain regions by many observers (Blench, 2012). Nevertheless, the wild donkey was hunted to the brink of extinction for its meat, and the consumption of equine meat is common in many Eurasian pastoral systems. In West Africa, the sale of donkeys for their meat has a long history; sick or exhausted animals that have been used for work in villages in semi-arid areas are slaughtered and their meat consumed (Blench, 2012). Due to this ambiguous status, the donkey trade remains poorly documented. However, it is reported in Burkina Faso that donkeys, in addition to their socio-economic and cultural importance, are slaughtered, and annual production was estimated at 800 tons according to official statistics from the Ministry of Animal Resources of Burkina Faso (MRA, 2008).

1.3.1.3 Nutritional Value

The biochemical composition of donkey milk is similar to that of human milk (Caroccio *et al.,* 2000). However, data on the biochemical and nutritional properties of donkey milk are limited. Table 1 shows the average composition per kilogram of cow's milk, donkey milk, mare's milk, and human milk.

Tableau 1: Average composition per kilogram of cow's, donkey's, mare's, and human 's milk.

Components	Cow	! Donkey	Mare	Female
Fat (g)	40	11	13	40
Protein (g)	34	17	21	19
Lactose (g)	48	66	64	65
Minerals (g)	7	4	4	2
Non-fat solids (g)	90	92	93	73
Total solids (g)	133	102	105	121
Cholesterol (mg)	140	22	45	200
Calcium (mg)	1200	680	890	320
Phosphorus (mg)	930	500	560	140

Saturated fatty acids (g)	24	4	4	18
Monounsaturated fatty acids (g)	11	2	3	16
Polyunsaturated fatty acids (g)	1	4	5	5

Source: Nikkhah (2012).

Table 1 shows that donkey milk contains more lactose and less fat than cow's milk; this high lactose content makes it suitable for the production of fermented products and beverages (Copola *et al.,* 2002). Donkey milk is similar to mare's milk, with a low total solids content (8%–10%), a low protein content (1.5%–1.8%), and a high lactose content (6%–7%). The fat content of donkey milk ranges from 0.28% to 1.82% (Chiavari *et al.,* 2005). The protein consists of 47.3% casein and 36.9% whey and contains approximately 30% β-lactoglobulin (Salimei *et al.,* 2004). It is also rich in amino acids and essential amino acids. Compared to all other milks, donkey milk is richer in valine and lysine (Nikkhah, 2012). Although most donkeys are used as working animals, there are reports indicating that under certain conditions, donkeys offer promising opportunities for milk production. For instance, China—the country with the world's largest donkey population—has recently improved donkey milk production through advanced breeding strategies, recording an annual output of 40,000 tons (Guo *et al.,* 2007; Yang *et al.,* 2006). It is commonly used in liquid form or as powdered milk, but there are also reports of efforts to make cheese.

Donkey meat can be considered a product of high nutritional value, characterized by a high content of unsaturated fatty acids and, more specifically, a large amount of polyunsaturated fatty acids (Polidori *et al.,* 2009), some of which play important roles as precursors of antithrombotic factors (Kinsella, 1988). These nutritional qualities of donkey meat are confirmed by a high content of essential amino acids, which is a very important factor in determining the quality of a food (Harper, 1999).

1.3.2 Importance of Donkey Breeding

1.3.2.1 Worldwide

- Global distribution of the donkey population

According to Bougler *et al.* (2005), the donkey population numbers approximately 43,472,000 head. China is the world's leading donkey producer, with nearly 11 million donkeys. Table 2 shows the global distribution of the donkey population.

Tableau 2: Global distribution of the donkey population

CONTINENTS	POPULATION (x1000)	Percentage (%)
Asia	14,885	34.24
Africa	9.639	22.17
Near and Middle East	9,220	21.21
Latin America and the	8,164	18.78

Caribbean		
Europe	1.512	3.48
North America	52	0.12
Total	**43.472**	100.00

Source: Bougler *et al.,* (2005)

Table 2 shows that Asia has the largest donkey population (over 14 million head), representing 34.24% of the global donkey population, followed by Africa with over 9 million, or 22.17% of the global population; North America has the smallest donkey population (less than 1 million). To explain these variations, one must compare the numerical figures for a continent with the role that the donkey plays in those regions.

- Donkey Population Trends

Table 3 shows donkey population trends by country in 1990.

Table 3: Donkey population trends by country worldwide in 1990.

WORLD		
Country	**Population (x1000)**	**Trend**
China	10,983	-
Ethiopia	5200	n
Pakistan	3,775	+
Mexico	3190	+
Iran	1900	+
India	1550	n
Egypt	1550	n
Brazil	1364	+
Afghanistan	1180	n
Nigeria	1000	+
Morocco	946	n
Turkey	895	-

+: increasing, -: decreasing, n: no change

Source: FAO (1994).

Table 3 shows that China and Ethiopia are the top two countries with the largest donkey populations; China has over 10 million, with a declining population, and Ethiopia has over 5 million, with numbers remaining stable.

1.3.2.2 In Africa

Table 4 shows the trend in donkey populations by country in 1990.

Table 4: Trends in donkey populations by country in Africa in 1990.

Country	**Population (x1000)**	**Trend**
Ethiopia	5,200	n
Egypt	1,550	n
Nigeria	1000	+

Morocco	946	n
Sudan	670	-
Mali	610	+
Niger	462	+
Burkina Faso	436	+
Senegal	364	+
Somalia	356	n
Algeria	340	n
Chad	271	n

+: increasing, -: decreasing, n: no change Source: FAO (1994).

Table 4 shows that Ethiopia has the largest donkey population in Africa, followed by Nigeria. In these two countries, the populations have remained unchanged. Populations are increasing in Nigeria, Mali, Niger, Burkina Faso, and Senegal, while Sudan is the only country experiencing a decline in its population.

1.3.2.3 In Cameroon

1.3.2.3.1 Population Trends

Table 5 shows the trends in livestock populations in Cameroon

Tableau 5: Changes in livestock populations (x1,000) in Cameroon from 1990 to 2000.

Years	**Donkeys**	Cattle	Goats	Sheep	Rabbits	Horses	Pigs	Chickens
1990	**35**	4,697	3,520	3,500	38	14.0	1,364	17,600
1991	**36**	4,830	3,550	3,550	39	15	1,370	18,000
1992	**36**	4,970	3,560	3,560	42	15.0	1,380	19,000
1993	**36**	5,110	3,760	3,770	44	15.0	1,390	20,000
1994	**36**	5,250	3,770	3,780	44	15.0	1,400	21,000
1995	**36**	5,400	3,800	3,800	45	15.5	1,410	22,000
1996	**37**	5,550	3,820	3,820	45	15.5	1415	22,000
1997	**37**	5,700	3,830	3,840	46	16.0	1,420	25,000
1998	**37**	5,900	3,840	3,860	46	16.0	1,425	28,000
1999	**37**	5,900	3,850	3,880	47	16.5	1,430	31,000
2,000	**38**	5,900	3,850	3,800	47	16.5	1,430	30,000

Source: MINEPIA (2003)

Table 5 shows that equine husbandry is relatively dominated by donkey production (38,000), with fluctuations from year to year.

1.3.2.3.2 Changes in Livestock Populations by Region from 2011 to 2012

Table 6 presents the evolution of camelid and equine herds in Cameroon by region from 2011 to 2012.

Tableau 6: Changes in Camellid and Equid Livestock Populations (in heads) in Cameroon by Region

	Donkeys		**Camels**	**Horses**		
Region	**2011**	**2012**	**2011**	**2012**	**2011**	**2012**

Adamaoua	8,727	8,937	-		4,149	2,675
Central	95	-			16	100
East	410	562	-	-	0	58
Far North	22,653	23,530	-	3	7,432	8,583
Coastal	-	-	-	-	-	758
North	14,426	16,105	26	37	1,817	1,897
Northwest	1,010	-			15,360	
West	251	140	-	0	441	455
South	2	2	3	3	15	15
Southwest	-	-	-	-	765	366
Total	**47,574**	**50,400**	29	43	30,045	16,468

Source: MINEPIA (2012)

Table 6 shows that donkey populations are concentrated in the Far North, North, Adamaoua, and Northwest regions and are increasing, while those of camels and horses are decreasing.

1.4 Major Donkey Breeds

1.4.1 Worldwide

There are many donkey breeds in the world. There are 185 recognized donkey breeds worldwide, distributed as follows: 51 in Europe, 26 in Africa, 32 in Asia and the Pacific, 24 in South America and the Caribbean, 47 in the Middle East, and 5 in North America (Siméon, 2008; Kugler *et al.*, 2008). These breeds vary in body weight, ranging from 80 to 480 kg, and in height at the withers, measuring 75 to 160 cm. Photos of some breeds are shown in Appendix 1.

In France: the Poitou donkey, the Grand Noir du Berry, the Provence donkey, the Cotentin donkey, the Norman donkey, the Pyrenean donkey, and the Bourbonnais donkey.

In Spain: the Catalan donkey, the Zamorano-Leones donkey, the Cordobes-Andaluz donkey.

In Italy: the Ragusano donkey, the Amiata donkey, the Asinara donkey, the Sardinian donkey, and the Martina Franca donkey. The Miranda donkey (Portugal), the Irish Pied donkey, the Teamster donkey (Australia), the Egyptian White donkey, the Abyssinian donkey (Ethiopia), the Miankala donkey (Mali), the Asino dell'Asinara (Italy), Criollo (Cuba), Asno Americana (Chile), Biyang (China), Black (Jordan), Dongolawl (Sudan), Tswana (Botswana), Bourik (Haiti), Kentucky Burro (Honduras, El Salvador, Brazil).

1.4.2 In Africa

1.4.2.1 Wild donkeys

These donkeys are primarily found in the arid regions of the Horn of Africa. Two subspecies are generally recognized: the Nubian and the Somali. Figure 1 shows the African wild ass.

Figure 1: **The African wild donkey.** Source: Siméon (2008)

- The Nubian donkey *(Equus asinus nubicus)*

It stands an average of 1.25 m at the withers. It has a gray coat, turning slightly reddish in summer, and bears a dark cross (or St. Andrew's cross) on the shoulder; the muzzle, belly, and inner sides of the legs are white. Its weight ranges from 200 to 240 kg. This donkey is believed to be the ancestor of our domestic donkeys. This breed, hunted for food but also for therapeutic purposes, is now endangered: only 1,500 to 3,000 individuals remain in Ethiopia and Sudan, countries where war and poaching are causing severe damage to the native wildlife. Figure 2 shows the Nubian donkey.

Figure 2: **The Nubian donkey** Source: Siméon (2008)

- The Somali donkey *(Equus asinus somaliensis)*

Its coat is gray, but tends toward mouse-gray. It stands 1.15 to 1.30 m at the withers and has abundant dark stripes on its legs. It is also distinguished by its black mane and the absence of a dorsal stripe and a crucial stripe.

It is hunted for its meat, and certain byproducts are listed in the pharmacopoeia. By 1992, only about 100 animals remained in the wild and approximately 70 donkeys were in captivity in Somalia. Since this population is too small to ensure the species' survival, these donkeys are unfortunately doomed to extinction (Siméon, 2008).

1.4.2.2 Domestic donkeys

Doutressoulle (1947) described six donkey breeds in French West Africa, which are:

> **Aïr Donkey**

With an average height of 1 m to 1.10 m, it has a gray and white or roan and white coat. It is fairly stocky and well-proportioned overall. The head is long and slender, with a narrow, short skull and a long muzzle. The neck is of medium length, the withers are powerful, the back is straight, and the croup is slightly sloping. This donkey inhabits the northern part of the Niger River bend. It is called Kobe by the local people.

> **Mauritanian Donkey**

Small in size, 0.90 m to 1.05 m tall, with short hair, a coat ranging from light gray to dark bay, and a distinct dorsal stripe. Square head, broad forehead, narrow nostrils, horizontal back, short croup, and clean-cut limbs.

> **Sahel Donkey**

Slightly larger and more slender than the Air donkey, it is more bony but muscular. The head is heavy and ungainly, with a narrow skull and a long face. Its coat is coarser than that of the Air donkey. It has a gray coat, sometimes lacking a dorsal stripe.

> **Gourma Donkey**

Of medium size, 1.05 m to 1.10 m, it has a gray coat with white predominating. Its lines are fairly harmonious; the body has a solid, well-built frame. Its habitat is the Niger River bend.

> **Minianka Donkey**

It is smaller, ranging from 0.90 m to 1.00 m, but stockier than the donkeys of the North. It is of average proportions but somewhat light overall. The head is long, with a straight muzzle, and the ears are long. The back is sturdy. It has a beige coat with a darker dorsal stripe and a darker midline. It is light-colored on the belly and limbs, with slight zebra-like markings on the lower legs, legs, and forearms.

> **Yatenga Donkey**

This is a strongly built, sturdy donkey of considerable size, ranging from 1.05 m to 1.15 m. The head is heavy and ungainly, with large ears. The muzzle is straight with a tendency toward a snub nose. The skeleton and musculature are more developed than in other varieties. Its coat is slate gray, sometimes tinged with black, with a very prominent dorsal stripe. The hair is fine, of medium length (3 to 5 cm), and the mane is quite thick. It is the Mossi donkey which, together with the Gourma donkey, constitutes the two best varieties.

1.5 Donkey Breeding Systems

Donkey management depends on the animal's purpose, herd size, intended use, and environment. Socioeconomic and cultural factors play an important role in donkey husbandry (Marshall and Ali, 2004; Mofya, 2004; Mrema, 2004).

1.5.1 Extensive system

Under this system, although limited by space, the donkey is allowed to graze freely throughout the day in the fields, drink, and rest wherever it pleases. These systems are common in the tropics, particularly under harsh conditions. During transhumance, when large herds of cattle move from northern Cameroon to the south in search of water and pasture, donkeys serve as a means of transport for children, women, and family luggage (Meutchieye *et al.,* 2012). After transporting goods from the fields or to the market, they are given supplements such as natron, salt, and cornmeal but generally do not receive veterinary care . The animals often have injuries caused by improper handling during transport or by fights with other animals.

1.5.2 Small-scale and mixed system

Donkeys are still largely considered a supplementary livestock resource for small-scale farmers in rural areas, where they are used as pack animals (Wod *et al.,* 2004; Yilmaz and Wilson, 2013). Pulling and transport are thus the main reasons for raising donkeys in a free-range system characterized by periodic rest and specialized tasks. This system is common throughout sub-Saharan Africa (Starkey, 1994).

1.5.3 Semi-intensive system

Here, the donkey or herd of donkeys is subject to periodic grazing and limited supplemental feeding at home, as well as spending part of the day in the fields and the entire night in the barn, pen, or shed. This can be observed on farms where donkeys have been integrated with other livestock, such as those in dairy production found in Kenya (Starkey, 1994). Fences are constructed with wood and wire to form paddocks that can be used to control grazing as a pasture management strategy.

1.5.4 Intensive system

Also known as zero-grazing, the intensive system involves one or a small number of donkeys that may be tethered or housed in a barn and fed in one location with sufficient quantities of hay, straw, vitamins, and concentrated feed. This can be seen in dairy herds, fattening herds, or with young animals (Polidori *et al.,* 2009). However, under traditional conditions, tethering is more a means of restriction or confinement than for a specific production purpose.

1.6 Zootechnical performance of donkeys

1.6.1 Growth performance

To the best of our knowledge, there is not enough information available regarding growth performance; however, Siméon (2008) notes that the birth weight of a foal is 25 kg.

1.6.2 Reproductive performance

Male and female donkeys become sexually active at one year of age, but in breeding programs, it is generally not recommended to breed donkeys before they are three years old. Estrous occurs in female donkeys on average every 21 days and lasts about a week. Under tropical conditions, female donkeys exhibit cyclical activity almost year-round, whereas in temperate countries, they breed seasonally (Blench *et al.,* 2004). Males are very active and can mount females several times a day. The gestation period for a female donkey averages 365 days. Blood circulation is established between the second and fifth months. In the fetus, weight gain becomes significant between the sixth and seventh months. By the ninth month, it reaches half of its birth weight. The final three months complete its development (Siméon, 2008), and birth weight is 25 kg. The age at first calving is 57 months, and the calving interval is 25.5 months in tropical countries (Blench *et al.,* 2004) and 3 years in temperate countries (Fielding, 1988).

1.6.3 Milk Production

A donkey produces approximately 1.85 kg of milk per day (Bordonaro *et al.,* 2013) for a donkey aged between 6 and 10 years. The general pattern of the lactation curve for female donkeys is similar to the standard pattern reported for major dairy ruminants, with peak production occurring 5 weeks after parturition (Bordonaro *et al.,* 2013). Younger female donkeys have a lower peak production but greater persistence.

1.7 Morphobiometric characteristics

In donkeys, the data most commonly collected for morphobiometric characterization include coat color, live weight (LW), chest circumference (CC), body length (BL), and height at the withers (HAW) (Ebangi and Vall, 1998).

1.7.1 Morphological characteristics

According to Raveneau and Daveze (1996), the donkey is distinguished by its fairly compressed chest, which is thicker and roughly cylindrical in shape toward the posterior. The back slopes slightly from the withers—which are more or less prominent—to the loin, which is straight and slightly depressed. The spine, shorter than that of a horse, consists of highly developed dorsal vertebrae, which is why donkeys have a prominent and strong back. The tail, set high, hangs down, is fairly long, and has a tuft of hair at its tip. The head is elongated, the muzzle white, the tip of the nose black, and the eyes almond-shaped. The pointed ears, set high, are set wide apart at the base with the ear flap directed forward; extremely mobile, they generally measure half the length of the head. Their limbs, as in all equines, are long, well-set apart from the body, and built for walking. The donkey's foot is straighter than that of a horse. This morphological characteristic must be kept in mind during trimming.

Their coat is generally gray except on the belly, muzzle, and around the eyes, which are white. Domestic breeds may be predominantly black, such as the Grand Noir de Berry, or brown, such as the Baudet du Poitou, or white, such as the Egyptian White Donkey. Gray breeds also have a distinctive band running down their back, known as the "St. Andrew's Cross."

Figure 3 illustrates the morphological characteristics of the donkey.

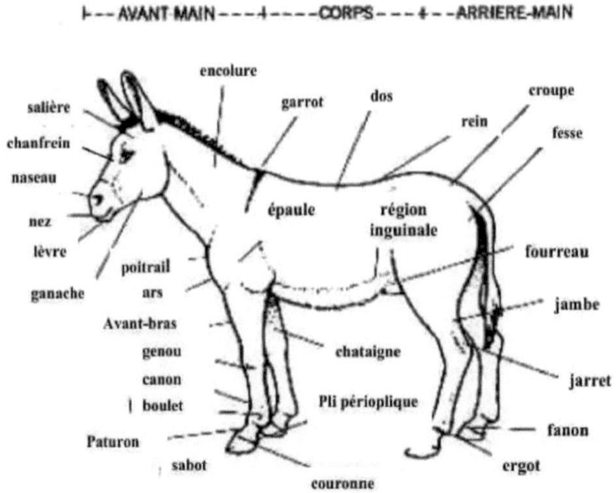

Figure 3: **Morphology of the donkey.** Source (asinerie.net, May 2015)

1.7.2 Main body measurements of the donkey

These morphobiometric parameters enable, on the one hand, the development of an equation for predicting live weight (Pearson and Ouassat, 1996; Ebangi and Vall, 1998) and, on the other hand, phenotypic characterization (Ebangi and Vall, 1998). Kefena *et al.* (2011), through the morphobiometric characterization of Ethiopian donkeys, confirmed the existence of six distinct donkey populations that had previously been misidentified. Ebangi and Vall (1998), meanwhile, were able to determine through morphobiometric characterization that donkeys from northern Cameroon are descended from *Equus asinus nubicus*.

a) Height at the withers

Height at the withers was the sole criterion used by Doutressoulle (1947) and Epstein (1971) to describe donkey resources in Africa. Sausseau (1925) found an average height at the withers for the Poitou donkey of 144 cm in males and 139 cm in females. Studies by Boujenane and Machmoum (2008a) recorded a withers height of 140.2, 139.7, and 138.3 cm, respectively, while Folch and Jordana (1997) found a withers height of 139.2 cm in Catalan donkeys. Ebangi and Vall (1998) obtained a withers height of 98.05 cm in donkeys from northern Cameroon. Salifou (2014) obtained a withers height of 105.1 cm in donkeys from Burkina Faso, and Roamba (2014) obtained a withers height of 108.9 cm in donkeys from Senegal.

b) Chest measurements

Chest dimensions indicate the metabolic potential of the subject under study (Bouchel, 1995; Bouchel et *al.*, 1997). The most common measurements of the metabolic cavity, as defined by Lauvergne (2007), are chest depth, chest circumference, and chest width.

The chest depth recorded in Littoral-Dinaric-type donkeys in Croatia is 42.03 cm (Ivankovic *et al.,* 2000) and that of donkeys in Albania is 46.1 cm (Papa *et al.,* 2012).

Ebangi and Vall (2005) recorded a chest circumference of 107.53 cm in donkeys from northern Cameroon; Roamba (2014) in Senegal recorded a value of 108.9 cm, and Salifou (2014) in Burkina Faso obtained 105.1 cm. Folch and Jordana (1997) obtained a chest circumference of 155.85 cm in Catalan donkeys in Spain, while Boujenane and Machmoum (2008a) found a chest circumference of 149.7 cm in the same donkeys and 152.4 cm in Poitevin donkeys in Morocco.

Kefena *et al.* (2011) recorded chest widths in Ethiopia for the Abyssinian, Afar, Hararghe, Ogaden, Omo, and Sinnar breeds of 21.97; 26.11; 25.88; 25.24; 26.55, and 24.60 cm, respectively. The chest width of Littoral-Dinaric donkeys in Croatia was 23.11 cm (Ivankovic *et al.,* 2000).

c) Body length

Ebangi and Vall (2005) recorded a body length of 110.19 cm in donkeys from northern Cameroon. Boujenane and Machmoum (2008a) recorded body lengths of 139.5 and 143.2 cm, respectively, in donkeys of the Catalan and Poitevine breeds.

d) Head dimensions

Folch and Jordana (1997) obtained a head length and width of 59.75 and 23.88 cm, respectively. Kefena *et al.* (2011) recorded head lengths in Ethiopia for the Abyssinian, Afar, Hararghe, Ogaden, Omo, and Sinnar breeds of 43.33, 45.39, 45.08, 45.97, 47.21, and 47.46 cm, respectively.

e) The Donkey's Skeleton

Each body dimension has a specific bony basis (Quittet, 1976). Skeletal evolution reflects the influence of the environment on animal metabolism (Salako, 2006). Figure 4 shows the skeleton of the donkey.

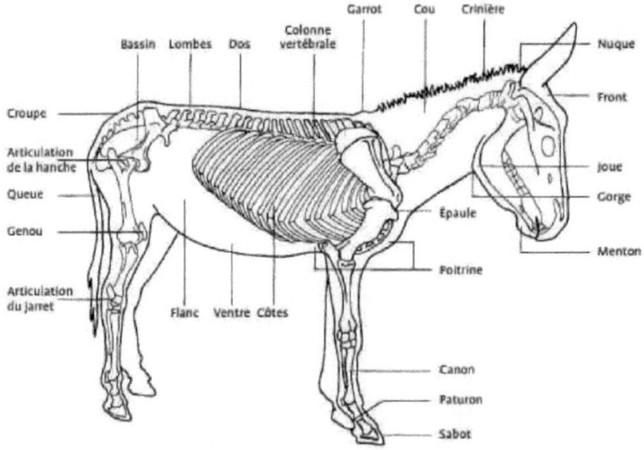

Figure 4: **The donkey's skeleton.** Source: Siméon (2008)

1.7.3 Biometric indices in donkeys

1.7.3.1 Definitions

Morphobiometric indices are ratios of body measurements used in the study of domestic animals (Bouchel, 1995). They reflect the proportions between the different parts of the donkey's body. There is a wide variety of morphobiometric indices used to assess donkey populations based on formulas described by several authors (Marcq *et al.,* 1951; Chabchoub *et al.,* 2004; *Nicks et al.,* 2006; Folch and Jordana, 1997; Boujenane *et al.,* 2008b).

1.7.3.2 Applications

The use of morphobiometric indices has made it possible to distinguish Barb horses and their derivatives in Algeria and Tunisia (Guedaoura *et al.,* 2011; Chabchoub *et al.,* 2004). Ebangi and Vall (2005) state that estimating morphobiometric parameters and the body profile index in donkeys can be a useful indicator for selecting draft donkeys to optimize their work capacity during the rainy season.

1.8 Barymetric Equations

Barymetric equations were developed as part of the monitoring of weight gain in large domestic livestock (Poivey *et al.,* 1980).

1.8.1 Foundations

Prediction equations (particularly for live weight) are based on correlations between body measurements (Bhadula *et al.,* 1979). A phase of slowed growth ( ) exists in all mammals, allowing for the development of valid prediction equations (Bouchel *et al.,* 1995). Correlations have been identified between morphology and certain production traits such as fleece (Litherland *et al.,* 1995). Kadifi *et al.* (2000) estimate that the correlation coefficients between chest circumference and live weight are the highest and most stable for various breeds raised in Europe.

1.8.2 Forms of Equations

There are many forms of predictive equations. Linear equations are the easiest to work with. Poivey *et al.* (1980) argue that it is possible to exploit the full range of mathematical relationships between dependent and independent variables. The mathematical models used in biometrics are, in fact, valid only under fairly homogeneous conditions (Bouchel, 1995). Multiple regressions can be used in research to predict live weight (Bhadula *et al.,* 1979). However, the use of chest circumference has yielded satisfactory results in donkeys according to several authors (Ebangi and Vall, 1998; Pearson and Ouassat, 2001; Salifou, 2014; Roamba, 2014); however, Ebangi and Vall (1998) state in their work that the introduction of a second variable (trunk length) did not significantly alter the value of the Pearson coefficient.

1.8.3 Barymetric Equations for Donkeys

Table 7 presents some barymetric equations for donkeys.

Table 7: Some barymetric equations for donkeys

Equation types	Equations	Authors
Linear equation	**$BW = 2.565 \times BT - 155.528$**	Salifou (2014)
	$PV = 2.035 \times PT + 0.805 \times LT - 182.702$	Roamba (2014)

Logarithmic equation	$\log(PV) = 2.548883 \times \log(PT) - 7.149$	Roamba (2014)
	$\log(PV) = 2.797 \times \log(PT) - 3.604$	Salifou (2014)
	$PV = PT^{(2.826)} / 4434.7$	Ebangi and Vall (1998)
Power equation	$PV = PT^{1,947} \times LT^{0,845} / 3834.8$	Ebangi and Vall (1998)
	$PV = PT^{(2.65)} / 2188$	Pearson and Ouassat (2001)

PV = live weight, PT = chest circumference, LT = trunk length

1.9 Population structure and phylogenetic relationships

1.9.1 Population structure

When individuals are described by a large number of variables, no simple graphical representation can be used to visualize the scatter plot formed by the data. Another method can be used that combines 57 multivariate analyses, such as principal component analysis (Cavalli-Sforza and Menozzi, 1994).

Principal Component Analysis (PCA) is a statistical method that can be used to reduce a large number of variables to a smaller number without losing too much information, and it provides a representation in a lower-dimensional space, thereby highlighting the variability within the population (Cavalli-Sforza and Menozzi, 1994). It is also used to highlight similarities or differences between variables and to identify the variables most strongly correlated with one another (Luanna et *al*., 2012).

1.9.2 Phylogenetic analysis

The primary objective of constructing a phylogenetic tree is to effectively describe the relationships between taxa in the living world (Carpentier, 2007). Methods based on similar classes are primarily used. As with goats, phylogenetic analysis is based on the use of quantitative traits to estimate similarities and distances between individuals (Carpentier, 2007).

One method for constructing a phylogenetic tree is the hierarchical ascending classification protocol, which aims to produce a simple schematic representation (Roux, 2006). The simplest objective of a classification is to divide the sample into groups of homogeneous observations, with each group being clearly distinct from the others (Carpentier, 2007).

CHAPTER II: MATERIALS AND METHODS

2.1 Study Area

2.1.1 Location and Altitude

The study was conducted in the highlands of northwestern Cameroon (Figure 5), which extend between latitudes$^{5^{\circ}}$ 45′ and$^{9^{\circ}}$ 9''N and longitudes 9° 13'' and 11° 13''E. It covers an area of approximately 17,400 km^2 at an altitude of 1,550 meters, dominated by a mountain range including the Oku Mountains, which rise to over 3,000 meters in elevation.

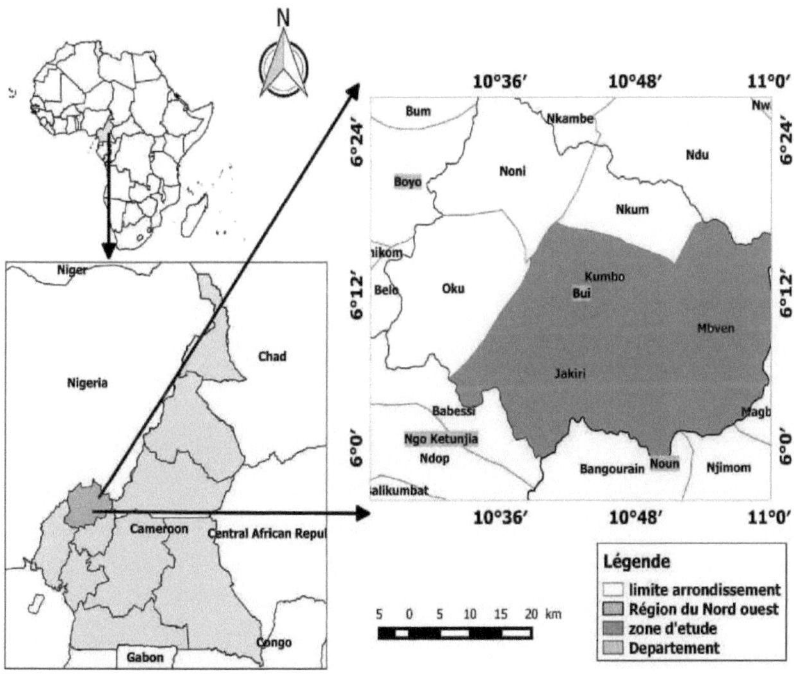

Figure 5: **Location of the study area**

2.1.2 Climate data

The climate is of the humid tropical type, with a relatively long rainy season (mid-March to late October) and a dry season that runs from November to March, with temperatures hovering around$^{22^{\circ}}$C; while minimum temperatures range between 13°C and 14°C, though values as low as 9 to 10.5°C are sometimes recorded on mountain peaks. November is the month with the lowest recorded temperatures. Precipitation ranges from 1,780 mm to 2,290 mm and can exceed 3,050 mm. The mountainous nature of this region is responsible for the variations in temperature s observed there.

2.1.3 Soils and Vegetation

The soils are generally ferralitic and lateritic, red or yellowish in color, with frequent occurrences of crusts and concretions. Superficial deposits of volcanic ash typically constitute the bulk of the soil in mountainous areas. The surface horizon is dark in color and very rich in organic matter, with a slow rate of decomposition and mineralization due to the climate (Cheek, 2000). P. africana grows in the Afro-montane and submontane zones, with the highest densities found in wooded savannas and secondary forests (NKENG, 2010). Thus, the main species associated with P. africana in the Bui department are the following: *Zanthoxylum rubescens, Cuviera longifolia, Entandrophragma angolense, Pouteria altissima, Strombosia scheffleri, Garcinia smeathmannii, Symphonia globulifera, Newtonia camerunensis, Alangium chinense, Ficus spp., Chassalia iaikomensis, Coffea liberica,* etc. There are also certain species resistant to bushfires, such as *Albizia gummifera, Gnidia glauca, Bridelia speciosa, and Croton macrostachyus.*

2.2 Animal material

2.2.1 Sampling

With the support of the regional office of the Ministry of Fisheries and Animal Industries (MINEPIA) in the Northwest, the various donkey-raising areas were identified. The departmental and district delegations facilitated contact with the various Zootechnical Centers for Sanitary and Veterinary Control (CZCSV) within their jurisdictions, and the officials served as liaisons with the breeders in their respective areas, explaining the purpose of the study. A sample of 174 adult donkeys (96 males and 78 females) was observed. The sample was distributed across 7 localities in the department of Bui. The criteria for selecting these localities focused on the presence of donkey farming and accessibility to these areas.

2.2.2 Selection of Animals

In the identified localities, 100 farms were selected. Data were also collected at three village markets, taking into account the animals' origin through interviews with the owners. Data were collected only on animals that appeared healthy and in good physical condition. Table 9 shows the sample of animals selected for the study by locality.

Table 8: Number of donkeys sampled by locality in the highlands of Northwest Cameroon.

Locations	Number	Percentage (%)
Vekovi	21	12.07
Lip	20	11.49
Reeh	30	17.24
Mumyu	22	12.64

Rifem	25	14.37
Njanawa	44	25.29
Mbouluf	12	6.90
Total	**174**	**100.00**

2.3 Data collection

Information on morphological characteristics was collected using a standardized questionnaire (FAO, 2013). For each animal, information on the location, rearing conditions, sex, physical description (coat color and other markings), and measurements (height at the withers, body length, chest circumference, neck length, ear length...) were recorded in a survey form (Appendix 3); the age of the animals, as reported by the breeders, was supplemented by dental examination as described in Appendix 4 (Daveze and Raveneau, 2002)

2.3.1. Morphology of donkeys from the highlands of northwestern Cameroon

Coat color was identified through direct observation in daylight, and the sex of the animal was determined.

2.3.2. Key measurements

Body measurements were obtained using the following equipment: a measuring tape for parameters related to lengths and circumferences, and a height rod for height parameters. Figure 6 presents the main measurements taken on the donkey.

Figure 6: **Measurements taken on donkeys.** Source: Kefena *et al.* (2011)

- Body length (BL): distance from the tip of the shoulder to the tip of the buttock.
- Croup length (Lcrp): the distance between the point of the hips and the tip of the buttocks.
- Withers height (WH): the distance from the highest point of the withers to just below the hoof

of the forelimb (this is the most frequently cited parameter for assessing the size of the animals).

- Rump height (RH): the distance from the highest point of the sacrum (between the ilium and the ischium) to the ground.
- Back height (BH): This is the distance from the middle of the back to the ground.
- Chest circumference (TP): or thoracic circumference (PT), measured by passing a measuring tape behind the withers at the point where the girths cross.
- Chest depth (PP): measured from the point where the girth passes to the withers-back boundary, or estimated at the girth line (behind the front legs).
- Chest width (LP): the distance behind the elbows.
- Croup width (lcrp): the distance between the two tips of the hips.
- Cannon bone circumference (CC): the circumference measured at the middle of the front cannon bone.
- Head length (LT): the distance from the nape of the neck to the point where the two nostrils meet.
- Ear length (EL) is measured from the outer edge, from the base to the tip.
- Neck length (LCou): the distance from the highest point of the withers to the nape of the neck.
- Head width (lt)
- Neck circumference (CCou)

2.3.3. Live weight

Live weight was estimated using the following equation: $LW = BW^{\times (2.826)} / 4434.7$ (Ebangi and Vall, 1998).

2.3.4. Biometric indices

Based on the various measurements, eight body indices were calculated using formulas described by several authors (Marcq *et al.,* 1951; Chabchoub *et al.,* 2004; Nicks *et al.,* 2006; Boujenane *et al.,* 2008b) developed for horses and adapted for donkeys (Folch and Jordana, 1997; Rezende *et al.,* 2013), namely:

Body Index or Ratio (ICor): chest circumference (PT) / height at the withers (HG). According to Folch and Jordana (1997), it reflects an animal's aptitude for work through its direct relationship to resistance to fatigue- , provided that the chest circumference does not exceed the withers height by more than 1/8, meaning that when this index is not greater than or equal to 1.125.

Body Profile Index (BPI): withers height (WH)/body length (BL); The body or corpulence index is used to evaluate an animal's proportions. A donkey is said to be short-bodied if the index is greater than 1, medium-bodied if it is equal to 1, and long-bodied for an index less than 1.

Compactness Index (CI): live weight (LW)/height at the withers (HAW);

Relative Body Index (RBI): body length (BL)/chest circumference (CC); this index reflects the animal's build; thus, an animal is considered long-bodied if RBI $\geq$90, medium-bodied if 84<RBI<90, and short-bodied if RBI $\leq$84.

Dactylothoracic Index (DTI): cannon bone circumference (CC) / chest circumference (CC); it reflects the relationship between body mass and the limbs that support it; this allows for the definition of three types: hypermetric (DTI > 11.5), eumetric (10.5 $\leq$ DTI $\leq$ 10.8), and elipometric or hypometric (DTI < 10.5).

Front-to-Rear Height (HDD): height at the withers (HG) / height at the croup (HC); this index reflects the animal's proportions. An animal is well-proportioned if this index equals 1.

Cephalic Index (ICeph): head width (It) / head length (LT). It reflects the proportions of the head and allows animals to be classified into two types: brachycephalic (ICeph > 100) and dolichocephalic (ICeph < 100).

Pelvic Index (IP): width of the rump (lcrp) / length of the rump (Lcrp). It provides an indication of the structure of the rump. A rump is well-proportioned if the width is approximately equal to the length; it is convex when IP < 1 and concave when IP > 1.

2.4 Statistical Analyses

Descriptive statistics were used to describe the distribution of coat color.

Analysis of variance (ANOVA) was used to test the influence of factors (sex, age, locality) on the various measurements and biometric indices. However, since the effects of sex and age were not significant, the following model was used: $y_{ij} = \mu + \alpha_i + e_{ij}$

Where μ: population mean;

α_i: fixed effect of location (i ranging from 1 to 7);

e_{ij}: error specific to each animal.

When the location factor was significant, Duncan's test was used to separate the means.

The direction and degree of association between traits were tested using Pearson's al correlation coefficient.

Barymetric equations were established by regressing live weight against the various measurements.

To assess the genetic variability of the population, Principal Component Analysis (PCA) was performed based on the 15 biometric parameters measured in the donkeys and live weight.

Discriminant factor analysis (DFA) based on the 15 body measurements and live weight (FAO, 2013) was performed to analyze the population structure.

The construction of the phylogenetic tree using the ascending hierarchical classification (AHC) protocol allowed us to establish the relationship between the genetic types.

All of these statistical analyses were performed using SPSS 21.0 and XLSTAT 2014 software.

CHAPTER III: RESULTS AND DISCUSSION

3.1 Results

3.1.1 Morphological characteristics of donkeys in the highlands of Northwest Cameroon

The coat color distribution among donkeys in the highlands of Northwest Cameroon is shown in Figure 7.

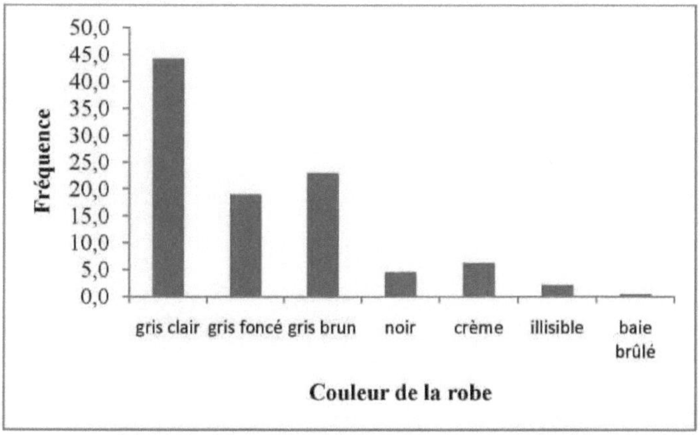

Figure 7: **Distribution of donkeys in the highlands of Northwest Cameroon by coat color.**

Figure 7 shows that the light gray coat is the most dominant (44.3%), followed by brownish gray (23.0%), dark gray (19.0%), cream (6.3%), black (4.6%), and an unidentifiable

(2.3%), and burnt bay.

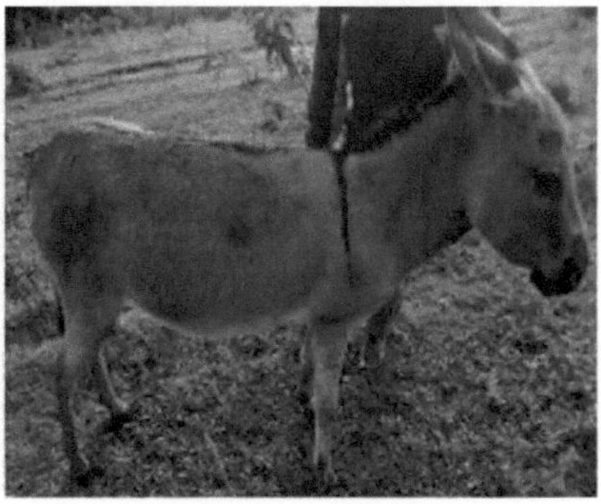

Dark gray coat **Light gray coat**

Black coat

Brown coat

Unreadable coat color

Burnt berry coat

Source: author

Figure 8: **Coat color of donkeys in the highlands of Northwest Cameroon**

3.1.2 . Main body measurements of donkeys from the highlands of Northwest Cameroon

3.1.2.1 Chest circumference, width, and depth of highland donkeys in Northwest Cameroon by locality

Analysis of variance (Appendix 2) shows that the three parameters were significantly influenced by locality ($P<0.05$).

Table 9 presents the means, standard deviations, and coefficients of variation for the various parameters of donkeys from the highlands of Northwest Cameroon by locality.

Table 9: Means, standard deviations, and coefficients of variation for chest circumference, chest width, and chest depth of highland donkeys in Northwest Cameroon by location.

Locations	N	**Chest circumference**	**Chest depth**	**Chest width**

		X ± s.d. (cm)	CV(%)	X ± s.d. (cm)	CV(%)	X ± s.d. (cm)	CV(%)
Age	21	110.22±5.96^{bc}	5.41	47.50 ± 3.9 ^{ab}	8.23	25.91±1.81^{ab}	0.98
Lip	20	111.07±4.59^{c}	4.13	47.45±2.61^{ab}	5.50	27.80±1.91^{c}	06.87
Reeh	30	107.00±5.43^{ab}	5.07	47.13±2.96^{a}	6.28	24.63±2.83^{a}	11.49
Mumyu	22	109.27±2.69^{bc}	2.46	48.68 ± 4.71 ^{ab}	9.67	27.27±2.05^{bc}	07.52
Rifem	25	108.88 ± 5.56 ^{abc}	5.11	52.36±3.47^{c}	6.63	25.11±1.95^{a}	07.76
Njanawa	44	108.41±4.84^{abc}	4.46	48.87±3.77^{ab}	7.72	25.68 ± 2.87 ^{a}	11.17
Mbouluf	12	105.71±5.14^{a}	4.86	49.67±4.03^{b}	8.11	24.67 ± 2.71^{a}	10.99
Total	**174**	**108.68±5.11**	**4.70**	**48.71±3.94**	**8.09**	**25.83±2.62**	**10.14**

a, b, c: Means marked with the same letter in the same column indicate that there is no significant difference (P > 0.05) between locations; X ± s.d.: mean ± standard deviation; CV: coefficient of variation.

- **Chest circumference**

Table 9 shows that the chest circumference averages 108.68 ± 5.11 cm with a coefficient of variation of 4.70%. Significantly higher means ($p < 0.05$) are observed in the localities of Mumyu, Vekovi, and Lip (109.27; 110.22, and 111.07 cm, respectively). Donkeys show greater variation in Rifem (CV=5.41%) and greater homogeneity in Mumyu (CV=2.46%) for this parameter.

- **Chest depth**

Chest depth averages 48.71±3.94 cm with a coefficient of variation of 8.09%. The chest of Rifem donkeys is deeper (52.36 cm) ($P<0.05$) than that of donkeys from other localities, and Mumyu donkeys show greater variability (CV=9.67%) for this parameter.

- **Chest width**

Chest width averages 25.83±2.62 cm with a CV of 10.14%. Donkeys from the Lip and Mumyu localities have a wider chest, and those from Reeh show greater variation (CV=11.49%).

3.1.2.2 Ear length, head length, and head width of donkeys from the northwestern highlands of Cameroon by locality.

Analysis of variance (Appendix 2) shows that ear length alone was not significantly influenced by locality ($P > 0.05$).

Table 10 presents the means, standard deviations, and coefficients of variation for ear length, head length, and head width of donkeys from the highlands of Northwest Cameroon, by locality.

Table 10: Means, standard deviations, and coefficients of variation for ear length, head length, and head width of donkeys from the highlands of Northwest Cameroon by location.

Locations		**Head width**		**Head length**		**Ear length**	
	N	X ± s.d. (cm)	CV (%)	X ± s.d. (cm)	CV (%)	X ± s.d. (cm)	CV (%)
Age	21	25.52±1.08^{b}	4.23	45.09±2.12^{cd}	4.70	25.38±1.75^{a}	6.89
Lip	20	24.60±1.05^{a}	4.27	42.90±2.2^{ab}	5.13	26.40 ± 1.67^{a}	6.33

Reeh	30	24.07±1.15[a]	4.78	42.18 ± 3.43[a]	8.13	25.78 ± 1.95[a]	7.56
Mumyu	22	23.95 ± 1.78[a]	7.43	43.61±2.16[abc]	4.95	26.31±1.60[a]	6.08
Rifem	25	24.21±1.44[a]	5.95	45.55 ± 2.28[d]	5.01	27.05 ± 1.51[a]	5.58
Njanawa	44	23.93±1.21[a]	5.06	43.84±2.17[bc]	4.95	26.49 ± 1.96[a]	7.40
Mbouluf	12	24.40±1.26[a]	5.16	43.17±2.29[a]	5.31	25.63±1.42[a]	5.54
Total	**174**	**24.30±1.36**	**5.60**	**43.74±2.65**	**6.06**	**26.21±1.81**	**6.91**

a, b, c, d: Means marked with the same letter in the same column indicate that there is no significant difference between the locations (p>0.01); X ± s.d.: mean ± standard deviation; CV: Coefficient of variation.

- **Head width**

Table 10 shows that the average head width of the subjects studied is 24.30±1.36 cm with a CV of 5.60%. The widest head (P<0.01) was found in Vekovi (25.52 cm). The highest coefficient of variation for this parameter was observed in the donkeys from Mumyu (7.43%).

- **Head length**

The average head length is 43.74 ± 2.65 cm with a coefficient of variation of 6.06%. The longest heads (P<0.01) were observed in Rifem and Vekovi (45.55 and 45.09 cm, respectively). In terms of variation, the donkeys from Reeh showed greater dispersion (CV=8.13%) and those from Vekovi were more homogeneous (CV=4.70%) for this parameter.

- **Ear length**

The average ear length is 26.21±1.81 cm with a coefficient of variation of 6.91%. This parameter is comparable across locations.

3.1.2.3 Body length, neck length, and neck circumference of donkeys from the highlands of northwestern Cameroon by locality.

Analysis of variance (Appendix 2) shows that of the three parameters, only body length was not significantly influenced by locality (P>0.05).

Table 11 presents the means, standard deviations, and coefficients of variation for body length, neck length, and neck circumference of donkeys in the highlands of Northwest Cameroon by locality.

Table 11: Means, standard deviations, and coefficients of variation for body length, neck length, and neck circumference of donkeys from the highlands of Northwest Cameroon by locality.

Locations		Body length		Neck length		Neck circumference	
	N	X ± s.d. (cm)	CV (%)	X ± s.d. (cm)	CV (%)	X ± s.d. (cm)	CV(%)
Age	21	107.67±6.17[a]	5.73	43.12 ± 2.92[a]	6.77	61.50 ± 4.17[b]	6.78
Lip	20	109.65±5.47[a]	4.99	43.75±2.77[a]	6.33	61.75 ± 4.03[b]	6.53
Reeh	30	110.10 ± 5.73[a]	5.21	42.97 ± 3.24[a]	7.54	56.49 ± 4.25[a]	7.52
Mumyu	22	110.77±6.22[a]	5.62	44.23 ± 2.93[a]	6.62	60.36 ± 3.97[b]	6.57
Rifem	25	111.95±7.13[a]	6.37	47.86 ± 2.30[b]	4.81	59.32 ± 3.43[b]	5.78

Njanawa	44	110.36±4.99[a]	4.52	42.09±3.21[a]	7.63	60.15 ± 3.62[b]	6.02
Mbouluf	12	108.13±7.47[a]	6.91	46.37±2.66[b]	5.74	54.71±4.32[a]	7.89
Total 174		**110.01±6.00**	**5.45**	**44.40±2.86**	**6.45**	**59.40±3.97**	**6.68**

a, b, c: Means marked with the same letter in the same column indicate that there is no significant difference (P>0.05) between locations; X ± s.d. (cm): mean ± standard deviation; CV: Coefficient of variation.

- **Body length**

Table 11 shows that the average body length is 110.01±6.00 cm with a CV of 5.45%, and this parameter is comparable across locations. Donkeys from Mbouluf show greater variability for this parameter.

- **Neck length**

The neck has an average length of 44.40 ± 2.86 cm with a coefficient of variation of 6.45%. Neck length is higher ($p<0.01$) in Rifem and Mbouluf (47.86 and 46.37 cm, respectively). Donkeys show greater variation in Njanawa (CV=7.63%) and greater homogeneity in Rifem (CV=4.81%) for this parameter.

- **Neck circumference**

The average neck circumference is 59.40 ± 3.97 cm, with a coefficient of variation of 10.94%. The lowest means ($P<0.05$) were found in Mbouluf and Reeh. Donkeys were less homogeneous in Mbouluf (CV=7.89%) and more homogeneous in Rifem (CV=5.78%) for this parameter.

3.1.2.4 Withers height, back length, and cannon bone circumference of donkeys from the highlands of northwestern Cameroon by locality.

Analysis of variance (Appendix 2) shows that of the three parameters, only back height was not significantly influenced by locality ($P > 0.05$).

Table 12 presents the means, standard deviations, and coefficients of variation for withers height, back height, and cannon bone circumference of donkeys from the highlands of Northwest Cameroon by locality.

Table 12: Means, standard deviations, and coefficients of variation for withers height, back height, and cannon bone circumference of donkeys from the highlands of Northwest Cameroon by locality

Locations		**Height at the withers**		**Back height**		**Cannon circumference**	
	N	X ± e.t (cm)	CV(%)	X ± s.d. (cm)	CV(%)	X ± s.d. (cm)	CV(%)
Age	21	100.50±4.43[b]	4.41	100.02 ± 5.27[a]	5.27	15.25±1.20[b]	07.87
Lip	20	100.60±2.96[b]	2.94	100.60±3.70[a]	3.68	15.37±1.22[b]	07.94
Reeh	30	98.00 ± 4.38[a]	4.47	97.87±4.08[a]	4.17	13.42±1.59[a]	11.84
Mumyu	22	100.23±2.64[ab]	2.63	98.82 ± 3.66 [a]	3.70	14.89 ± 1.98 [b]	13.29
Rifem	25	100.96 ± 3.46[b]	3.43	100.44 ± 3.23[a]	3.21	16.52 ± 1.91[c]	11.56
Njanawa	44	99.16±3.11[ab]	3.14	98.41±3.78[a]	3.84	14.65±1.76[b]	12.01

Mbouluf	12	100.04±2.62^{ab}	2.62	99.08±2.47^{a}	2.49	15.58 ± 1.31 ^{bc}	08.41
Total	**174**	**99.72±3.57**	**3.58**	**99.13±3.80**	**3.83**	**14.93±1.85**	**12.39**

a, b, c: Means denoted by the same letter within the same column indicate that there is no significant difference between the locations (P>0.05); X ± s.d.: mean ± standard deviation; CV: Coefficient of variation.

- **Height at the withers**

Table 12 shows that the average height at the withers is 99.72±3.57 cm with a coefficient of variation of 3.58%. The highest means (P<0.05) are recorded in Rifem (100.96 cm), Lip (100.60 cm), and Vekovi (100.50 cm); the Reeh donkeys show greater variation (CV = 4.47%), while those from Mbouluf are more homogeneous (CV = 2.62%) for this parameter.

- **Withers height**

Withers height averages 99.13±3.80 cm with a coefficient of variation of 3.83%. This parameter is comparable across locations (P>0.05).

- **Cannon circumference**

The average is 14.93±1.85 cm with a coefficient of variation of 12.39%. The lowest mean (P<0.05) was recorded among donkeys from Reeh (13.42 cm) and the highest in Rifem (16.52 cm), while donkeys from Mumyu showed greater variation (13.29%) for this parameter.

3.1.2.5 Length, width, and height of the rump of donkeys from the highlands of northwestern Cameroon by locality.

Analysis of variance (Appendix 2) shows that none of the three parameters was significantly influenced by locality (P > 0.05).

Table 13 presents the means, standard deviations, and coefficients of variation for rump measurements in donkeys from the highlands of Northwest Cameroon by locality.

Table 13: Means, standard deviations, and coefficients of variation for the length, width, and height of the rump of the local donkey from the highlands of northwestern Cameroon, by locality.

Locations	N	Rump length X ± e.t (cm)	CV(%)	Rump width X ± s.d. (cm)	CV(%)	Rump height X ± s.d. (cm)	CV (%)
Age	21	33.97±1.88	5.53	30.83±2.85	9.24	103.36±4.09	3.96
Lip	20	34.63±2.27	6.56	31.80±2.57	8.08	104.40±2.91	2.79
Reeh	30	33.58±1.53	4.56	30.77±1.72	5.59	102.52±4.14	4.04
Mumyu	22	33.73±1.54	4.57	30.97±1.70	5.49	103.11±4.28	4.15
Rifem	25	34.27±1.45	4.23	31.50±1.44	4.57	104.55±3.84	3.67
Njanawa	44	33.68±1.76	5.23	31.28±2.16	6.91	102.63±3.98	3.88
Mbouluf	12	32.83±2.08	6.34	30.42±2.11	6.94	103.75±3.44	3.32
Total	**174**	**33.83±1.78**	**5.26**	**31.13±2.11**	**6.78**	**103.29±3.91**	**3.78**

a, b, c: Means marked with the same letter in the same column indicate that there is no between locations (P>0.05); X ± s.d.: mean ± standard deviation; CV: coefficient of variation.

- **Rump Length**

Table 13 shows that the average rump length of the local donkey is 33.83±1.78 cm, with a coefficient of variation of 5.26%, and is comparable across locations. The donkeys from Rifem are more homogeneous (CV=4.23%), while those from Lip are more dispersed (CV=6.56%).

- **Rump width**

The width of the rump averages 31.13 ± 2.11 cm, with a coefficient of variation of 6.78%, and is comparable across locations. Donkeys from Vekovi show greater variation (CV = 9.24%), while those from Rifem are more homogeneous (CV = 4.57%) for this parameter.

- **Height at the rump**

The average height at the rump is 103.29±3.91 cm with a coefficient of variation of 3.78% and is comparable across locations. Donkeys from Lip are more homogeneous (CV=2.79%) and those from Mumyu more dispersed (CV=4.15%).

3.1.2.6. Live weight of donkeys from the highlands of Northwest Cameroon by locality

Analysis of variance (Appendix 2) shows that live weight was significantly influenced by locality (P<0.05).

Table 14 presents the mean, standard deviation, and coefficient of variation of the live weight of donkeys in the highlands of Northwest Cameroon by locality.

Table 14: Mean, standard deviation, and coefficient of variation of live weight of donkeys in the highlands of Northwest Cameroon by locality

Locations		**Live weight**	
	N	$\bar{X}$ ± s.d. (kg)	CV (%)
Age	21	134.15±20.38^{bc}	15.19
Lip	20	136.73±15.67^{c}	11.46
Reeh	30	123.30±17.87^{ab}	14.49
Mumyu	22	130.20 ± 9.05 ^{abc}	6.95
Rifem	25	129.53±18.10^{abc}	13.97
Njanawa	44	127.76±16.35^{abc}	12.79
Mbouluf	12	119.05±17.02^{a}	14.29
Total	**174**	**128.68±17.07**	**13.26**

a, b, c: Means denoted by the same letter within the same column indicate that there is no significant difference between the locations (P > 0.05); X ± s.d.: mean ± standard deviation; CV: coefficient of variation.

Table 14 shows that the average live weight of adult donkeys in the highlands of northwestern Cameroon is 128.68 ± 17.07 kg. The highest weights (P<0.05) were recorded in Lip and Vekovi (136.73 and 134.15 kg, respectively), and the lowest in Mbouluf (119.05 kg). The donkeys from Reeh showed greater variation (CV=14.49%), while those from Mumyu were more homogeneous (CV=6.95%).

3.1.2.7 Correlation between body measurements of donkeys in the highlands of Northwest Cameroon.

Table 15 illustrates the types of correlations between body measurements in donkeys from the highlands of Northwest

.

Tableau 15: Correlations between body measurements in donkeys from the highlands of Northwest Cameroon

	PV	LC	HG	PT	LCou	CC	LT	It	LO	PP	LP	lcrp	Lcrp	HC	HD	PC
PV	1															
LC	0,58**	1														
HG	0,64**	0,59**	1													
PT	**0,99****	0,59**	0,64**	1												
LCou	0,30**	0,49**	0,52**	0,30**	1											
CC	0,55**	0,31**	0,48**	0,56**	**0,15**	1										
LT	0,47**	0,41**	0,56**	0,47**	0,43**	0,35**	1									
It	0,42**	0,37**	0,46**	0,41**	0,27**	0,35**	0,34**	1								
LO	0,36**	0,39**	0,38**	0,37**	0,23**	0,24**	0,23**	**0,13**	1							
PP	0,45**	0,57**	0,57**	0,45**	0,61**	0,26**	0,45**	0,35**	0,28**	1						
LP	0,48**	0,46**	0,50**	0,49**	0,28**	0,44**	0,30**	0,41**	0,19*	0,37**	1					
lcrp	0,64**	0,48**	0,46**	0,63**	0,26**	0,34**	0,37**	0,27**	0,45**	0,31**	0,34**	1				
Lcrp	0,67**	0,53**	0,53**	0,67**	0,24**	0,40**	0,41**	0,42**	0,41**	0,37**	0,44**	0,78**	1			
HC	0,58**	0,60**	0,82**	0,59**	0,48**	0,36**	0,47**	0,47**	0,37**	0,57**	0,45**	0,44**	0,53**	1		
HD	0,62**	0,57**	0,88**	0,62**	0,53**	0,41**	0,50**	0,48**	0,29**	0,55**	0,46**	0,41**	0,48**	0,83**	1	
PC	0,31**	0,39**	0,54**	0,31**	0,58**	0,32**	0,47**	0,31**	0,17*	0,53**	0,30**	0,25**	0,25**	0,45**	0,51**	1

** .P< 0.01.* P< 0.05 (two-tailed)

HG: height at the withers; **LC**: body length; **PT**: chest circumference; **LO:** ear length; **HD**: height at the back; **HC:** height at the rump; **LCrp**: rump length; **lcrp:** rump width; **LCou**: neck length; **PC**: cannon bone circumference; **CCou**: neck circumference; **LT**: head length; **IT**: head width; **Lp**: chest ; **PP**: chest depth; **PV**: live weight.

Table 15 shows that all measurements taken are positively correlated at the 5% or 1% significance level, except for neck length and neck circumference, ear length, and head width (P>0.05). The highest correlation coefficients are found between live weight and chest circumference (r=0.99), wither height and back height (r = 0.88), wither height and rump height (r = 0.82), and rump height and back height (r = 0.83), whereas the lowest correlation coefficient was found between ear length and head width (r = 0.13).

3.1.3 Body weight prediction equations for donkeys in the highlands of Northwest Cameroon

Table 16 illustrates the various forms of equations for predicting the live weight of donkeys in the Northwest region of Cameroon based on body measurements.

Tableau 16: Types of weight prediction equations based on body measurements

Types of Equations	Body measurements	Equations	R^2
Linear	PT	PV = 3.34 PT - 233.89	0.99
	LCrp	PV = 6.39 Lcrp + 87.59	0.45
Exponential	PT	$PV=7.52e^{0.026}$ PT	0.99
	LCrp	$PV = 23.73e^{\wedge 0.049Lcrp}$	0.45
Logarithmic	PT	PV = 360.6ln (PT) - 1561	0.99
	LCrp	PV = 216.1 LCrp + 632.1	0.45
Power	PT	$PV = 0.001(PT)^{2.826}$	1
	LCrp	$PV = 0.338\ LCrp^{1.68}$	0.45
Polynomial	PT	$PV = 0.028PT^2 + 2.760(PT) + 96.81$	1
	LCrp	$PV = 0.15\ LCrp^2 - 3.84\ LCrp + 80.26$	0.45

PT: chest circumference; BW: body weight; LCrp: rump length; R^2: coefficient of determination

Table 16 shows that chest circumference is a good predictor of live weight in donkeys from the highlands of Northwest Cameroon (R^2 ranging from 0.99 to 1). The linear equation using chest circumference, due to its simplicity and high coefficient of determination, appears to be better suited for predicting live weight.

3.1.4 Biometric indices of donkeys from the highlands of Northwest Cameroon by locality

3.1.4.1. Body Index, Profile Body Index, and Relative Body Index

Analysis of variance (Appendix 2) shows that the three biometric indices were significantly influenced by locality ($P<0.01$).

Table 17 presents the means, standard deviations, and coefficients of variation for the body index, profile body index, and relative body index of donkeys from the highlands of Northwest Cameroon by locality

Tableau 17: Means, standard deviations, and coefficients of variation for body index, the Profil Body- , and the relative body index of donkeys from the highlands of northwestern Cameroon by locality

Locations	ICOR		ICP	ICR		
	$\bar{X} \pm$ s.d.	CV(%)	$\bar{X} \pm$ s.d.	CV (%)	$\bar{X} \pm$ s.d.	CV (%)
Age	1.09 ± 0.05^b	4.58	0.94 ± 0.05^d	5.32	0.98 ± 0.05^a	5.10
Lip	1.11 ± 0.03^b	2.70	0.92 ± 0.03^{bcd}	3.26	0.99 ± 0.03^a	3.03
Reeh	1.09 ± 0.04^b	3.67	0.90 ± 0.03^a	5.33	1.03 ± 0.04^b	3.88
Mumyu	1.09 ± 0.03^b	2.75	0.91 ± 0.04^{abc}	4.39	1.01 ± 0.05^b	4.95

Rifem	1.08±0.04^{ab}	3.70	0.90 ± 0.04^{ab}	4.44	1.03±0.05^{b}	4.85
Njanawa	1.09±0.04^{b}	3.67	0.90±0.04^{ab}	4.44	1.02±0.04^{b}	3.92
Mbouluf	1.06±0.04^{a}	3.77	0.93±0.06^{cd}	6.45	1.01±0.06^{b}	5.94
Total (174)	**1.09±0.04**	**3.67**	**0.91±0.04**	**4.39**	**1.01±0.05**	**4.95**

a, b, c: Means marked with the same letter in the same column indicate that there is no significant difference between the locations (P > 0.05); X ± s.d.: mean ± standard deviation; CV: coefficient of variation; ICOR: Body Condition Score; ICP: Profile Body Condition Score; ICR: Relative Body Condition Score.

Table 17 shows that:

- The body index of donkeys in the highlands of Northwest Cameroon averages 1.09 ± 0.04 cm with a CV of 3.67%. The lowest mean (P<0.05) is found in Mbouluf (1.06). These results indicate that donkeys from the highlands of Northwest Cameroon are suitable for work due to their direct relationship to fatigue.
- The body index for profile is on average 0.91 with a CV of 4.39%; the highest index (P<0.05) was recorded in Vekovi (0.94). These results suggest that the donkeys of the Northwest Highlands are of a slender build. The donkeys of Mbouluf are more dispersed (CV=6.45%), while those of Lip (CV=3.26%) are more homogeneous.
- The relative body index averages 1.01 with a CV of 4.95%. The lowest indices are recorded in Vekovi and Lip (0.98 and 0.99, respectively). These results allow these animals to be classified as long-bodied.

3.1.4.2 Compactness Index, Dactylothoracic Index, Front-to-Back Height

Analysis of variance (Appendix 2) shows that of the three indices, only the front-to-back height was not significantly influenced by location (P > 0.05).

Table 18 presents the means, standard deviations, and coefficients of variation for the compactness index, dactylothoracic index, and front-to-back height index of donkeys from the highlands of northwestern Cameroon by locality.

Table 18: Means, standard deviations, and coefficients of variation for the compactness index, dactylothoracic index, and front-to-back height index of donkeys from the highlands of Northwest Cameroon ( ) by locality.

Locations	CI		IDT		HDD	
	— X ± e.t	CV (%)	— X ± e.t	CV (%)	— X ± e.t	CV (%)
Age	1.33±0.17^{bc}	12.78	0.14±0.01^{bc}	7.14	0.97±0.03^{a}	3.09
Lip	1.36 ± 0.13^{c}	09.56	0.14 ± 0.01^{bc}	7.14	0.96 ± 0.02^{a}	2.08
Reeh	1.25±0.14^{ab}	11.20	0.13±0.01^{a}	7.69	0.96±0.02^{a}	2.08
Mumyu	1.30±0.08^{bc}	0.15	0.14±0.02^{b}	14.28	0.97±0.02^{a}	2.06
Rifem	1.27±0.15^{abc}	11.81	0.15±0.02^{d}	13.33	0.97±0.01^{a}	1.03
Njanawa	1.28 ± 0.14^{bc}	10.94	0.14±0.02^{b}	14.28	0.97 ± 0.02^{a}	2.06

Mbouluf	1.19±0.15[a]	12.61	0.15±0.01[cd]	6.67	0.96±0.01[a]	1.04
Total (N=174)	**1.29±0.15**	**11.63**	**0.14±0.02**	**14.28**	**0.97±0.02**	**2.06**

a, b, c: Means marked with the same letter in the same column indicate that there is no significant difference between locations (P>0.05); X ± s.d.: mean ± standard deviation; CV: coefficient of variation; CI: compactness index; IDT: dactylothoracic index; HDD: front-to-back height.

Table 18 shows that:

- The compactness index averages 1.29 ± 0.15 with a CV of 11.63%. The highest indices (P < 0.05) are recorded in Lip and Vekovi (1.36 and 1.28, respectively), while the lowest index is recorded in Mbouluf (1.19). Furthermore, the donkeys in Vekovi are more dispersed (CV=12.78%), while those in Mumyu are more homogeneous (CV=6.15%).

- The dactylothoracic index averages 0.14 ± 0.02 with a CV of 14.28%. These results allow the donkeys of the highlands of northwestern Cameroon to be classified as hypermetric. The lowest significant value was recorded in Reeh (0.13), and the highest in Rifem (0.15) and Njanawa (0.15). Donkeys are more homogeneous in Mbouluf (CV = 6.67%) and more dispersed (CV = 14.28%) in Mumyu and Njanawa.

- The front-to-back height index averages 0.97±0.02 with a CV of 2.06%. These results show that donkeys from the highlands of Northwest Cameroon have a greater height at the rump than at the withers. The localities are comparable for this parameter (P>0.05).

3.1.4.3 Cephalic and Pelvic Indices

Analysis of variance (Appendix 2) shows that of the two biometric indices, only the cephalic index was significantly influenced by locality (P<0.01).

Table 19 presents the means, standard deviations, and coefficients of variation for the cephalic index and pelvic index of donkeys from the highlands of Northwest Cameroon by locality.

Table 19: Means, standard deviations, and coefficients of variation for the cephalic index and pelvic index of donkeys from the highlands of Northwest Cameroon by locality

Locations	Cep Index	PI		
	— X ± s.d.	CV (%)	— X ± s.d.	CV (%)
Age	0.57±0.03[bc]	5.26	0.91±0.05[a]	5.49
Lip	0.57±0.03[c]	5.26	0.92±0.04[a]	4.35
Reeh	0.57±0.04[c]	7.02	0.92±0.03[a]	3.26
Mumyu	0.55±0.04[ab]	7.27	0.92±0.04[a]	4.35
Rifem	0.53 ± 0.03[a]	5.66	0.92±0.03[a]	3.26
Njanawa	0.55 ± 0.03[ab]	5.45	0.93±0.04[a]	4.30
Mbouluf	0.57 ± 0.03 [bc]	5.26	0.93±0.03[a]	3.23
Total (N=174)	**0.56±0.03**	**5.88**	**0.92±0.04**	**4.35**

a, b, c: Means marked with the same letter in the same column indicate that there is no significant difference between the locations (P>0.05); X ± s.d.: mean ± standard deviation; CV: coefficient of variation. ICeph: cephalic index; IP: pelvic index.

Table 19 shows that:

- The cephalic index averages 0.56±0.03. These results suggest that donkeys from the highlands of northwestern Cameroon belong to the dolichocephalic donkey group. The lowest index is recorded (P<0.05) in Rifem (0.53), and the donkeys from Mumyu show greater variation (CV=7.27%) for this parameter.

- The pelvic index of the donkeys averages 0.92±0.04. These results suggest that the rump of donkeys in the highlands of Northwest Cameroon is convex. The localities are comparable for this parameter.

3.1.4.4 Correlation between biometric indices of donkeys from the highlands of Northwest Cameroon

Table 20 presents the correlation coefficients between the various biometric indices in donkeys from the highlands of Northwest Cameroon.

Tableau 20: Correlations between biometric indices of donkeys from the highlands of Northwest Cameroon.

	ICOR	ICP	ICR	IC	IDT	HDD	Iceph	IP
ICOR	1							
ICP	-0,35**	1						
ICR	-0,44**	-0,70**	1					
IC	**0,81****	-0,26**	-0,37**	1				
IDT	-0,40**	0,04	0,27**	-0,18*	1			
HDD	-0,18*	0,26**	-0,11	-0,03	0,08	1		
Iceph	0,02	-0,02	0,01	-0,06	-0,16*	-0,15	1	
IP	0,15	-0,05	-0,05	0,18*	0,02	0,04	-0,16*	1

ICOR: Body Index or Body Ratio; ICP: Body Profile Index; HDD: Height at Withers; IDT: Dactylothoracic Index; ICR: Relative Body Index; IC: Compactness Index; Iceph: Cephalic Index; IP: Pelvic Index.

Table 20 shows that the correlation coefficients vary widely. They can be negative or positive depending on the pair of indices. A number of coefficients are not significant.

20.1.5. Genetic variability, structure, and genetic relationships among donkey populations in the highlands of northwestern Cameroon.

20.1.5.1.Genetic variability of the population

Principal component analysis was performed based on 16 quantitative variables, namely: withers height, back length, body length, chest circumference, neck circumference and length, head length and width, chest depth and width, ear length, rump height, rump length and width, cannon bone circumference, and live weight.

Table 21 illustrates the contribution of the 16 principal components to the genetic variability observed within the population.

Tableau 21: Eigenvalue and cumulative variance of the different principal components in the analysis of the variability observed within the donkey population of the highlands of northwestern Cameroon.

Components (SE)	Variables	Value Eigen	Variance %	Variance cumulative
CP 1	Live weight (LW)	7.92	49.48	49.48
CP 2	Body length (BL)	1.64	10.28	59.76
CP 3	Height at the withers (HG)	1.07	6.70	66.46
CP 4	Chest circumference (CC)	0.77	4.81	71.27
CP 5	Neck length (Nl)	0.71	4.41	75.68
CP 6	Neck circumference (Ncirc)	0.68	4.24	79.92
CP 7	Head length (HL)	0.64	4.02	83.94
CP 8	Head width (lT)	0.52	3.24	87.18
CP 9	Ear length (LO)	0.49	3.06	90.24
CP10	Chest depth (CD)	0.38	2.37	92.60
CP11	Chest width (CW)	0.36	2.28	94.88
CP12	Rump width (lcrp)	0.35	2.16	97.04
CP13	Rump length (LCrp)	0.19	1.19	98.23
CP14	Height at the croup (HC)	0.17	1.07	99.30
CP15	Back height (HD)	0.11	0.68	99.98
CP16	Barrel circumference (PC)	0.01	0.01	100.00

Table 21 shows that the first 9 components account for 90.24% of the observed variability within the population. However, it should be noted that the first two components (live weight and body length) account for 59.76% of the variance observed within our sample, with live weight contributing 49.48% and body length 10.28%. For this reason, the population structure will be analyzed using these first two components.

3.1.5.2 Population Structure

Discriminant factor analysis (DFA) revealed that the study population consists of three genetic types. Table 22 shows the biometric characteristics of these three genetic types, and Figure 9 illustrates the structure of the donkey population.

Tableau 22: Biometric characteristics of the three genetic types

Types	PV	LC	HG	PT	LCou	CC	LT	lt	LO	PP	LP	lcrp	Lcrp	HC	HD	PC

A	134.02	111.71	100.79	110.42	45.03	60,51	44.05	24.52	26.69	49.55	26.78	31.47	34.32	104.47	100.13	15.26
B	111.13	105.84	97.00	103.30	43.12	56.64	42.52	23.65	25.18	46.70	23.99	29.72	32.42	100.44	96.36	14.19
C	164.05	116.00	103.61	118.61	45.43	63.54	46.78	25.39	26.85	52.21	26.93	34.14	36.14	106.71	103.43	15.94

WG: withers height; BL: body length; CH: chest circumference; EL: ear length; HD: height at the back; **HC:** height at the croup; **LCrp**: length of the croup; **lcrp:** width of the croup; **LCou**: length of the neck; **PC**: cannon bone circumference; **CCou**: neck circumference; **LT**: length of the head; **IT**: width of the head; **Lp**: chest width; **PP**: chest depth; **PV**: live weight.

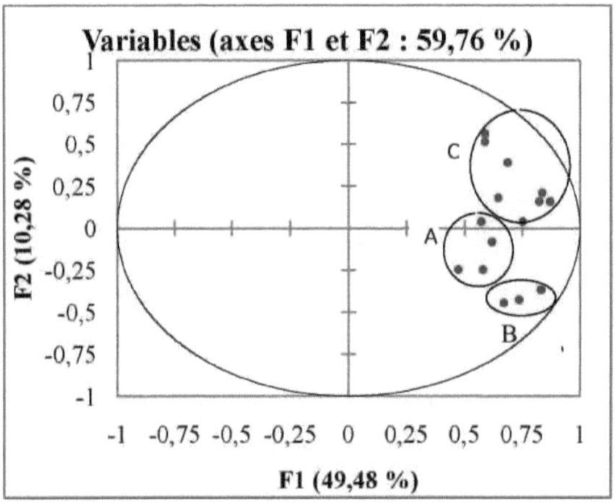

Figure 9: **Structure of the donkey population in the Northwest.**

Figure 9 shows that the donkey population of the highlands of northwestern Cameroon consists of three genetic types—A, B, and C—whose biometric characteristics are listed in Table 22. The table indicates that genetic type C has high characteristics, type A has average characteristics, and type B has low characteristics. The observed differences may be due to various endogenous factors on the one hand (species, breed, strain, etc.) and exogenous factors on the other (climate, forage availability, husbandry practices, and the animals' origin).

3.1.5.3. Genetic relationships among the genetic types of the highlands of Northwest Cameroon

The phylogenetic tree in Figure 10 illustrates the genetic relationships among the three genetic types of the highlands of Northwest Cameroon.

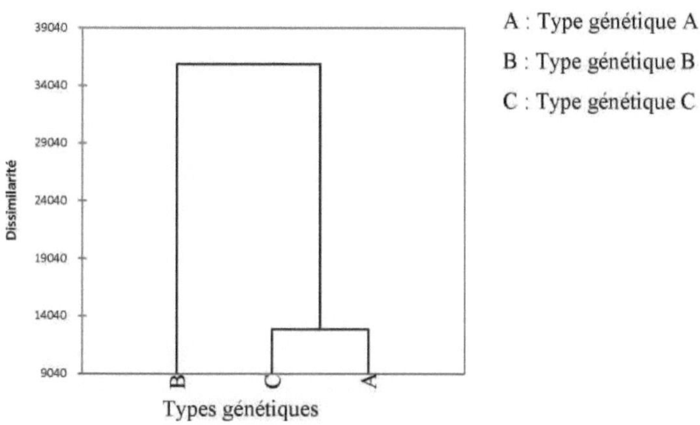

Figure 10: **Phylogenetic tree of genetic types in the highlands of Northwest Cameroon.**

Figure 10 shows that the phylogenetic tree, while confirming the population structure, has made it possible to establish the relationships that exist between the genetic types. Thus, genetic types A and C are more closely related, whereas genetic type B is more distantly related. This closeness is likely linked to the genetic distance that exists between the three genetic types. Furthermore, the population appears to consist of two subgroups: one comprising genetic types A and C, and the second consisting solely of genetic type B.

3.2 Discussion

3.2.1 Morphology

The predominant coat color consists of shades of gray (86.2%) with a black stripe on the shoulder and white markings around the nostrils, the ventral region, and the inner part of the legs; this description corresponds to that provided by Raveneau and Daveze (1996) for *Equus asinus nubicus* or *Equus asinus africanus*, which originates from the Horn of Africa north of Sudan. Similar observations regarding coat color have been made by several authors (Ebangi and Vall, 1998; Papa and Kume, 2012).

3.2.2 Body measurements

The chest circumference (110.07 cm), although relatively high in the donkeys of our sample, does not differ significantly from that recorded in donkeys from northern Cameroon, which is 107.53 cm (Ebangi and Vall, 2005). The same applies to the results recorded by Roamba (2014) in Senegal (108.9 cm) and by Salifou (2014) in Burkina Faso (105.1 cm).

The chest depth (48.80 cm) recorded in our study is relatively greater th t that recorded in Littoral-Dinaric-type donkeys in Croatia, which is 42.03 cm (Ivankovic *et al.,* 2000), and does not differ significantly from that recorded in Albania, which is 46.1 cm (Papa and Kume, 2012).

The chest width (25.85 cm) of the donkeys in our sample, although greater than that of the Abyssinian breed (21.97 cm) in Ethiopia (Kefena *et al.,* 2011) and that of the Littoral-Dinaric type in Croatia (Ivankovic *et al.,* 2000) (23.11 cm), is not significantly different from that of the Hararghe (25.88 cm), Ogaden (25.24 cm), and Sinnar (24.60 cm) breeds in Ethiopia (Kefena *et al.,* 2011).

Our results regarding head width (24.29 cm) are higher than those obtained by Folch and Jordana (1997) for the Catalan donkey (23.43 cm) in Spain.

Kefena *et al.* (2011) found higher head length values in the Afar (45.39 cm) and Omo (47.21 cm) breeds in Ethiopia than those in our results (43.75 cm). However, our results are higher than those obtained for the Catalan breed, which was 39.36 cm (Folch and Jordana, 1997).

Our results regarding ear length (26.21 cm) are similar to those obtained by Roamba (2014) in Senegal (26.5 cm) and by Salifou (2014) in Burkina Faso (24.6 cm) and are lower than those obtained for the Catalan donkey (33.13 cm) in Spain (Folch and Jordana, 1997).

Roamba (2014) and Salifou (2014) found body length values that were relatively lower than our results in Senegal (105.5 cm) and Burkina Faso (103.9 cm), respectively. In contrast, our results are close to those of donkeys from northern Cameroon (110.19 cm).

The neck length (44.41 cm) we recorded is higher than that obtained for Abyssinian-type donkeys (43.33 cm) and lower than that of the Omo (47.21 cm) and Sinnar (47.46 cm) types (Kefena et *al.,* 2011).

The average neck circumference of the local donkey is 59.44 cm. Such measurements have not yet been reported.

Wilson (1981) reports that there is little physical variation among donkeys across Africa and that they rarely exceed 105 cm in height at the withers. This is confirmed by our results (99.73 cm), which are below 105 cm.

On average, the back height of the subjects studied is 99.12 cm, which is close to the height at the withers. These results do not corroborate those of Kefena *et al.* (2011), who found that donkeys in Ethiopia are taller at the back than at the withers.

Our results regarding cannon bone circumference (14.95 cm) are higher than those obtained by Nengomasha (1997) for donkeys in Zimbabwe (14 cm). However, they are lower than those obtained for Catalan donkeys (18.73 cm) in Spain (Folch and Jordana, 1997).

Folch and Jordana (1997) found relatively higher values for rump length (45.01 cm) than those in our results (33.82 cm) in Spain.

Kefena *et al.* (2011) and Folch and Jordana (1997) found values for rump width higher than those in our results (31.09 cm) in Ethiopia (33.78 cm) and Spain (42.68 cm), respectively.

Our results regarding height at the rump (103.28 cm) are higher than those obtained by Ivankovic *et al.* (2000) for the Littoral-Dinaric donkey (99.85 cm) in Croatia. Similarly, Kefena *et al.* (2011) obtained lower values for rump height (95.42 cm) in the Abyssinian donkey than those in our results. The results obtained by several authors (Ivankovic *et al.,* 2000; Kefena *et al.,* 2011; Folch and Jordana, 1997) show that donkeys are taller at the rump than at the withers. This is confirmed by our results.

Salifou (2014) found live weight values (114.02 kg) in Burkina Faso that were lower than those in our results (128.69 kg), as did Roamba (2014) in Senegal and Ebangi and Vall (2005) in northern Cameroon (123.8 and 123.45 kg, respectively). These differences may be due to the rearing system and the type of use. Indeed, the animals in our study area are used for transport, which is a less strenuous activity than that to which draft animals are subjected in the Northern Cameroon region, where only 6% are used for transport.

Of all the parameters studied, locality did not significantly influence ($P>0.05$) body length, ear length, rump length and width, height at the rump, and height at the withers.

The high variability observed in chest width and depth, body weight, and cannon bone circumference may be of great interest in the future when genetic improvement objectives are considered. This high variability may result from the different husbandry systems among the populations, as some animals graze freely while others are tethered. Environmental factors may also be responsible for these varying degrees of variability observed within the population.

3.2.3 Correlation between body measurements

De Aluja *et al.* (2005), as well as Pearson and Ouassat (1996); Nengomasha *et* al. (1997) found a high correlation coefficient between live weight and chest circumference. This corroborates our results. However, this correlation coefficient is equal to that obtained by De Aluja *et al.* (2005) and higher than that of Pearson and Ouassat (1996); Nengomasha *et al.* (1997). A high correlation coefficient between withers height and croup height was obtained by Nicks *et al.* (2006) in horses, which corroborates our results.

3.2.4 Equations for predicting live weight

Chest circumference proved to be the best predictor of live weight ($R^2= 0.99$). This is consistent with the results of several authors (Nengomasha *et al.*, 1997; Pearson and Ouassat, 1996; Pearson and Ouassat, 2001).

3.2.5 Biometric indices

The ICor values we recorded are close to those obtained by Folch and Jordana (1997) for the Catalan breed in Spain.

As for the ICP, we obtained a value of 0.91; this characterizes the donkeys of the highlands of northwestern Cameroon as long-bodied, as the value is less than 1. These results corroborate those of Folch and Jordana (1997). Meanwhile, Guedaoura *et al.* (2011) obtained an index greater than 1 for Barb horses and their derivatives.

The relative body index value obtained (1.01) allows the donkeys of the highlands of Northwest Cameroon to be classified as long-bodied; this morphology was confirmed by the body index, which indicates the degree of compactness of the chest circumference and classifies the animals as long-bodied, medium-bodied, and short-bodied, showing that these animals are lean and slender.

The compactness index of the donkey studied showed little variation. No references regarding this index were reported.

Folch and Jordana (1997) recorded IDT values lower than those in our study but both above 11.5; this allows them to be classified as hypermetric-type animals.

Our results regarding the front-to-back height (HDD) suggest that donkeys from the highlands of northwestern Cameroon have a larger posterior region than the anterior region. These results do not corroborate those of Folch and Jordana (1997) and Garcia (2006), who found that the Catalan breed is well-proportioned.

Garcia (2006) and Folch and Jordana (1997) obtained cephalic index values for Spanish breeds that were lower than those in our study. However, both the Spanish breeds and our study population are of the dolichocephalic type.

The IP values we recorded are relatively lower than those reported by Folch and Jordana (1997) for the Catalan donkey breed and by Garcia (2006) for the Spanish breeds (Andalusian, Catalan, Zamorano-Leonese). However, this index for the donkey population in our study and that of the Spanish breeds is less than 1, indicating that the rump has a concave structure.

The positive correlation coefficient between ICOR and IC within the donkey population of the highlands of northwestern Cameroon is r = 0.81. Such results have not yet been reported.

3.2.6 Genetic Variability, Structure, and Genetic Relationships Among Genetic Types 3.2.6.1 Genetic Variability

Principal component analysis showed that of the 16 components, the first 9 components account for 90.24% of the observed variability within the population. However, it should be noted that the first two components (live weight and body length) account for 59.76% of the variance observed within our sample, with 49.48% (live weight) and 10.28% (body length). Garcia (2006) found that the three main components account for 99.71% of the total variability among donkey breeds in

Spain, and of the 26 variables collected, only 10 were most effective in distinguishing the different breeds in Spain. However, live weight alone accounts for 72.20% of the total variability among donkey breeds in Spain.

3.2.6.2 Population Structure

AFD analysis indicates that there are three genetic types of donkeys in the highlands of northwestern Cameroon. Roamba (2014) found that the donkey population in Senegal is homogeneous, as did Salifou (2014) in Burkina Faso. However, Papa and Kume (2012) identified two genetic types in Albania.

3.2.6.3 Genetic relationships among genetic types

The phylogenetic tree constructed using the hierarchical ascending classification protocol revealed the existence of three genetic types, with two closely related types (types A and C) and one more distantly related to the other two, type B.

CONCLUSION

At the conclusion of our study, which aimed to analyze the morphobiometric characteristics of the donkey population in the Northwest Highlands, the following findings emerged:

- There is significant variability in coat color, with a predominance of gray and its variants (light gray, dark gray, brownish gray) and a scarcity of burnt bay and indistinct colors;

- The withers height, chest circumference, body length, croup height, and live weight of donkeys in the Northwest Highlands are, respectively: 99.72±3.57 cm; 108.68 ± 5.11 cm; 110.01 ± 6.00 cm; 103.29 ± 3.91 cm; 128.68 ± 17.07 kg;

- The correlations between the measurements are highly variable (positive, negative, high, or low);

- Chest circumference is a good predictor of weight in donkeys in our study area. The simple barometric equation PV = 3.34 PT – 233.89 (R^2= 0.99), with a higher and nearly perfect coefficient of determination, may be a better predictor of live weight.

- Biometric indices are also highly variable, revealing animals that are long-bodied, hypermetric, with large thoracic capacity, and dolichocephalic faces;

- Principal component analysis revealed high genetic variability within the donkey population in our study area, and discriminant analysis showed that our population consists of three genetic types: Type A with average characteristics, Type B with low characteristics, and Type C with high characteristics.

- The dendrogram confirmed the existence of these three genetic types, grouped into two subgroups: subgroup 1, consisting of genetic types A and C, which are very closely related, and subgroup 2, consisting of genetic type B, which is more distantly related.

- There is considerable phenotypic variability among donkeys in the highlands of northwestern Cameroon with respect to all the traits studied. This diversity suggests the possibility of genetic improvement of the donkey species using traditional methods of selection and crossbreeding.

OUTLOOK

It is advisable to continue and expand genetic analyses to increase productivity, particularly through informed selection. Thus, we plan to:

1. Completing the inventory and monitoring of donkey resources based on qualitative and zootechnical criteria; in addition to refining the national donkey map, this will help identify the effects of farming systems on donkey performance across different agroecological regions;

2. Expanding studies on donkey biodiversity across all agro-ecological zones of Cameroon;

3. Identify multiple predictors of live weight in different age categories of the local donkey in Northwest Cameroon; this will involve further research into body composition in Cameroonian donkeys;

4. Estimate body composition and carcass yields in donkeys from Northwest Cameroon; this would improve meat production;

5. Establish the biochemical profile of Cameroonian donkeys based on biochemical and chromosomal markers; this would lead to the reconstruction of the phylogeny of Cameroonian donkeys.

BIBLIOGRAPHICAL REFERENCES

Aganga, A. A., Tsopito, C. M., Seabo, D. 1994. Donkey power in rural transportation. Botswana case study. *Appropriate Technology Journal, **21**:* 32-33.

Albano, B. P., Phillip, R. E., Nuno, F., S. J., Amel O. B., Mohammed, A., Abdalla, M., Mashkour, J., Jordana, P. T., Gordon, L. 2004. African Origins of the Domestic Donkey. *Science, **304**.*

Aubineau, M., Bermond, A., Bougler, J., Ney, B., Roger-Estrade, J. 2002. Larousse agricole.

Audiot, A. 1995. Breeds of Yesterday for the Livestock Industry of Tomorrow, Paris, INRA Press.

Bertoni, C. 2000. Genetic Contribution to the Conservation of the Poitou Donkey. Thesis. National School of Agricultural Engineering. Bordeaux, France. Animal Production, Aquaculture, and Poultry Science Track.

Bhadula, S. K., Bhat, P. N., Garg, R. C. 1979. Prediction of body weight from body measurements in sheep. *Indian Journal of Animal Sciences*, ***49*** (10): 775-777.

Blench, R. 2012. Wild asses and donkeys in Africa: interdisciplinary evidence for their biogeography, history, and current use. (http://www.rogerblench.info/RBOP.htm).

Blench, R. M., de Jode, A., Gherzi, E. 2004. Donkeys in Nigeria: history, distribution, and productivity. Donkeys, people, and development. A resource book of the Animal Traction Network for Eastern and Southern Africa (ATNESA). Starkey P and Fielding, D Editions.

Bogart, R. 1965. Modern Methods of Livestock Improvement. Paris, France. Intercontinental Organization eds. New York. 8vo.

Bordonaro, S., Dimauro, C., Criscione, A., Marletta, D., Macciotta, N. P. P. 2013. The mathematical modeling of the lactation curve for dairy traits of the donkey (*Equus asinus*). *Journal of Dairy Science*, ***96***: 4005-4014.

Bouchel, D. 1995. Contribution to the study of morpho-biometric indices and their conditions of use for the mapping of caprine genetic resources. Diploma in Introduction to Scientific Research, François Rabelais University, Tours, France.

Bouchel, D., Lauvergne, J.J., Guibert, E., Minvielle, F. 1997. Morphobiometric study of the Rove goat. Withers height, chest depth, substernal space, and substernal gracility index in females. *INRA Productions Animales* **148,** (6): 37-46.

Bougler, J., Duplan, J-M., Zongo, D. 2005. Characteristics and global distribution of animal species exploited by humans. Manual of Comparative Animal Breeding, Thewis et al. Editions.

Boujenane, I., Machmoum, M. 2008a. Body measurements of Poitevine, , and Catalan donkeys and their crosses in Morocco. *Journal of Animal Husbandry and Veterinary Medicine in Tropical Countries,* **61** (1): 63-67

Boujenane, I., Touati, I., Machmoum, M. 2008b. Body measurements of Arabian-Barb horses in Morocco. *Journal of Veterinary Medicine,* **159**: 144-149.

Boumaza S. 1974. Herd Management: Collection of Technical Documentation for Trainers. Algiers, Algeria. Ministry of Agriculture and Agrarian Reform, Directorate of Agricultural Education.

Carpentier, F. G. 2007. Introduction to Multidimensional Analysis. France. Editions Cheminements.

Carroccio, A., Cavataio, F., Montaldo, G.D., Amico, D., Alabrese, L., Iacono, G. 2000. Intolerance to hydrolyzed cow's milk protein in infants: clinical characteristics and dietary treatment. *Clinical and Experimental Allergy,* **30**: 1597-1603.

Cavalli-Sforza, L. L., Menozzi, P. 1994. The History and Geography of Human Genes. Princeton University Press. Princeton, New Jersey: 203–205.

Chabchoub, A., Landolsi, F., Jary, Y. 2004. Study of the morphological parameters of Barb horses in Tunisia. *Journal of Veterinary Medicine*, **155**: 31–37.

Chappez, G. 2000. The Donkey: History, Myth, and Reality. France: *Cabédita.*

Cheek, M., Onana, J., Pollard, B. 2000. The Plants of Mount Oku and the Ijim Ridge, Cameroon: A Conservation Checklist. Kew, Royal Botanic Gardens.

Chiavari, C., Coloretti, F., Nanni, M., Sorrentino, E., Grazia, L. 2005. Use of donkey's milk for a fermented beverage with lactobacilli. *Lait*, **85**: 481-490

Coppola, R., Salimei, E., Succi, M., Sorrentino, M., Nanni, M., Ranieri, P., Belli, R., Blanes, J., Grazia, L. 2002. Behavior of *Lactobacillus rhamnosus* strains in donkey's milk. *Annals of Microbiology*. **52**: 55-60.

Danchin, B. C. 2002. Local goat breeds. Institut d'élevage.

Darwin, C. 1859. On the Origin of Species by Means of Natural Selection, or the Preservation of Favoured Races in the Struggle for Life. London, Murray. French translation of the *6th edition* (Murray, London, 1872) by Edmond Barbie: The Origin of Species by Means of Natural Selection, or the Struggle for Existence in Nature.

Daveze, J., Raveneau, A. 2002. The Book of the Donkey. France: Editions Rustica.

De Aluja, A. S., Pérez, T. G., Lopez, R., Pearson, R. A. 2005. Liveweight estimation of donkeys s in Central Mexico from measurement of thoracic circumference. *Tropical Animal Health and Production*, **37** (1): 159-171.

De Rechambeau, H., Verrier, E., Bidamel, J. P., Maignel, L., Labroue, F., Tribout, T., Palhier, T., Astruc, J., Barille, S., Chapui, H. 2003. Implementation of procedures for monitoring genetic variability in selected domestic animal populations and establishment of management guidelines: Application to dairy sheep and pigs, Proceedings of the BRG.

De Rochambeau, H. 1990. Objectives and methods of genetic management of rabbit populations with limited numbers. Mediterranean option. Seminar series no. 8: 1927.

Doutressoulle, G. 1947. Livestock farming in French West Africa. Paris. Editions Larose.

Ebangi, A. L., Vall, E. 1998. Phenotypic characterization of draft donkeys within the Sudano-Sahelian zone of Cameroon. *Journal of Animal Husbandry and Veterinary Medicine in* Tropical *Countries*, **51**: 327-334.

Ebangi, A. L., Vall, E. 2005. Dry season effect on live weight and some body dimensions of working donkeys in the Sudano-Sahel region of Cameroon. *Tropicultura.* **23** (1): 48-52.

Epstein, H. 1971. The Origin of Domestic Animals in Africa. New York: Africana Publishing Corporation, vol. 2.

Fadlaoui, A. 2006. Bioeconomic Modeling of Animal Genetic Resource Conservation. Doctoral dissertation in Agricultural Science and Biological Engineering. Catholic University of Louvain. Belgium.

Fanica, P. O. 2008. Milk, Cows, and City Dwellers: From the 17th to the 20th Century, Quae Publishing.

Farissier, S., 2007. The Donkey. France. Artemis Publishing.

Fielding, D., 1988. Reproductive characteristics of the jenny donkey (*Equus asinus*): A Review. *Tropical Animal Health and Production*, **20**: 160-166.

Flamant, J. C., 1988. The Human Dimension of Genetic Improvement Programs for Sheep Breeds. Biology and Animals. Presses de l'Institut d'études politiques de Toulouse. France.

Folch, P., Jordana, J., 1997. Characterization, reference ranges, and the influence of gender on morphological parameters of the endangered Catalan donkey breed. *Journal of Equine Veterinary Science,* **17** (2): 102-111.

Food and Agriculture Organization (FAO) 1994. FAO Production Yearbook. Statistics Series 117: Rome, Italy. FAO.

Food and Agriculture Organization (FAO) 1997. Design, execution, and analysis of livestock breed- -surveys: a case study in Zimbabwe.

Food and Agriculture Organization (FAO) 2011. Molecular genetic characterization of animal genetic resources. Rome, Italy. FAO Animal Production and Health Guidelines. No. 9.

Food and Agriculture Organization (FAO) 2013. Phenotypic characterization of animal genetic resources. Rome. FAO Animal Production and Health Guidelines No. 11.

Garcia, E. M. 2006. Morphological, hematological, and clinical biochemical characterization of five Spanish donkey breeds for conservation programs. Thesis submitted in partial fulfillment of the requirements for the degree of Doctor of Veterinary Medicine. University of Barcelona. Faculty of Veterinary Medicine. Department of Animal Science and Food. Unit of Genetics and Animal Breeding.

Groves, C. P. 1966. Taxonomy. On the African Wild Ass. C.P. Groves, F. Ziccardi, and A. Toschi Editions.

Groves, C. P. 1986. The Taxonomy: Distribution and Adaptations of Recent Equids. Equids in the Ancient World, eds. R.H. Meadow and H.P. Uerpmann.

Guedaoura, S., Cabaraux, J. F., Moumene, A., Tahraoui, A., Nicks, B. 2011. Morphometric evaluation of Barb horses and their derivatives in Algeria. *Annals of Veterinary Medicine,* **155**: 14-22.

Guo, H. Y., Pang, K., Zhang, X. Y., Zhao, L., Chen, S. W., Dong, M. L., Ren, F. Z. 2007 Composition, physicochemical properties, nitrogen fraction distribution, and amino acid profile of donkey milk. *Journal of Dairy Science,* **90:** 1635-1643

Habault, P. 1974. Elements of General Animal Science. Paris: Editions J-B Baillière.

Haltenorth, T., Diller, H. 1980. A field guide to the mammals of Africa, including Madagascar. London: Collins Editions.

Harper, A. E. 1999. Defining the essentiality of nutrients. In M. E. Shills, J. A. Olson, M. Shike, & A. C. Ross (Eds.), Modern nutrition in health and disease (9th ed.). Baltimore: Williams and Wilkins Ed.

Helmer, D. 1979. Research on the food economy and the origin of domestic animals, based on a study of post-Paleolithic mammals (from the Mesolithic to the Bronze Age) in Provence. Doctoral dissertation. University of Science and Technology. Languedoc. Montpellier.

National Institute of Statistics (INS) 2013. Statistical Yearbook of Cameroon, Chapter 15: Livestock and Fisheries. 265–282.

Ivankovic, A., Caput, P., Mioc, B., Pavic, V., 2000. The phenotype features of donkeys in Croatia. *Agriculturae conspectus scientificus,* **65** (2): 99–105.

Jean-Pierre, D. 2009. Man and Domestic Animals: Anthropology of a Passion. Fayard.

Jussiau, R., Montméas, L., Parot, J.-C. 1999. Animal Husbandry in France: 10,000 Years of History, Educagri Editions.

Kadifi, N., Baillargeon, J., Pelletier, R., Mellor, S., Brnalochu, E., Lapointe, J., Sneeck, L., Baldassare, H. 2000. Use of body weight in goats. In proceedings of the 7thInternational Conference on Goats, France.

Kefena, E., Beja-Pereira, A., Han, J. L., Haile, A., Mohammed, Y.K., Dessie, T. 2011. Eco-geographical structuring and morphological diversities in Ethiopian donkey populations. *Livestock Science* **141**: 232-241.

Kinsella, J. E. 1988. Food lipids and fatty acids: Importance in food quality, nutrition, and health. *Food Technology,* **42**: 124-145.

Kugler, W., Grunenfelder, H. P., Broxham, E. 2008. Donkey Breeds in Europe. Inventory, Description, Need for Action, Conservation. Report 2007/2008. Monitoring Institute for Rare Breeds and Seeds in Europe & Save Foundation.

Lauvergne, J. J. 2007**.** Case for a study of animal genetic resources (AGR) in an African country. In ed.

Lauvergne, J. J., Bourzart, D., Minville, F. 2000. Using morphobiometric indices to map goat resources in Africa. In R.M. Blench, K. C. MacDonald (eds.): *The origin and development of African livestock: archaeology, genetics, linguistics, and ethnography*, UCL Press, London and New York. Pp. 290–301.

Lauvie, A. 2007**.** Managing small local animal populations: an approach to the diversity of implemented measures. *Doctoral thesis,* AgroParisTech. Management of animal genetic resources.

Litherland, A. J., Dollé, P. D. J., O'Neill, K. T. 1995. Length as a predictor of live weight rank in New Zealand goats. New Zealand *Journal of Agricultural Research*. **38**: 361–365.

Luanna, C. P. 2012. Biometric variability of goat populations revealed by means of principal component analysis. *Genetics and Molecular Biology*. **35, (**4): 777-782.

Luurt, O. 2001. Use of donkeys for traction and plowing. Agrodok 35.

Machado, T., Lauvergne, J. J., Zafindrajaona, P. 1992. The Brazilian goat population: The scenario since its discovery. *Archivos de Zootecnia*. **41** (154): 455-466.

Mahaman S. Z. 1986. Cattle, Sheep, and Goat Farming in Niger: An Ethological Study. Inter-State School of Veterinary Sciences and Medicine (EISMV).

Marcq, J., Lahaye, J., Cordiez, E. 1951. The Horse's Conformation. Gembloux:[3rd]edition. Duculot.

Marshall, K., Ali, Z. 2004. Gender Issues in Donkey Use in Rural Ethiopia. Donkeys, People, and Development: A Resource Book for the Animal Traction Network for Eastern and Southern Africa. Wageningen, Netherlands: Fielding and Starkey Editions.

Meutchieye, F., Kwalar, N. N., Nyock, R. A. F. 2012. Current Donkey Production & Functionality Chapter 12 Donkey Husbandry and Production Systems.

Ministry of Livestock, Fisheries, and Animal Industries (MINEPIA) 2003. National Report on the Zoogenetic Resources of Livestock in Cameroon. Yaoundé, Cameroon. Animal Genetic Resources Service.

Ministry of Livestock, Fisheries, and Animal Industries (MINEPIA) 2012. Activity Report. Yaoundé, Cameroon. Animal Genetic Resources Service.

Ministry of Animal Resources of Burkina Faso (MRA) 2008. Statistics on the Livestock Sector in Burkina Faso. Ouagadougou, Burkina Faso. MRA.

Mofya, R. 2004. Social consequences of introducing donkeys into Zambia. Donkeys, people and development: A resource book for the Animal Traction Network for Eastern and Southern Africa. Wageningen, Netherlands. Fielding and Starkey edition.

Mrema, M. 2004. Economic and gender issues of donkey use in Kweneng and Kgatleng Districts, Botswana. Donkeys, People, and Development: A Resource Book for the Animal Traction Network for Eastern and Southern Africa. Wageningen, Netherlands. Fielding and Starkey edition.

Nengomasha, E. M., Jele, N., Pearson, R. A. 1997. Morphological characteristics of working donkeys in southwestern Zimbabwe. Debre Zeit, Ethiopia. Animal Traction Network for Eastern and Southern Africa: Report of the international ATNESA workshop held May 5–9, 1997.

Nicks, B., Delfontaine, B., Canart, B., Vanderbruggen, J., Vandenheede, M. 2006 Morphological characteristics of Belgian Draft mares. *Annales de Médecine Vétérinaire,* **150:** 247-251.

Nikkhah, A. 2012. Equidae milk promises substitutes for cow and human breast milk. *Turkish Journal of Veterinary and Animal Sciences*, **36** (5): 470-475

Nkeng, F. P., Verina I., Awono, A. 2010. Assessment of *Prunus africana* bark exploitation methods and sustainable exploitation in the South West, North-West, and Adamaoua regions of Cameroon, CIFOR, Yaoundé, Cameroon.

Ouragh, L., Amigue, S. Y., Nguyen, T. C., Bosher, M. Y. 2002. Genetic analysis of Moroccan sheep breeds. *Renc. Rech. Ruminants*. **9**: 99.

Papa, L., Kume, K. 2012. Results of the identification and characterization of the donkey population in Albania. *Agriculture and Forestry,* **58** (3): 125-134.

Pearson, R. A., Ouassat, M. 1996. Estimation of live weight and body condition of working donkeys in Morocco. *Veterinary Records*. **138**: 229-233.

Pearson, R.A., Ouassat, M. 2001. A guide to live weight estimation and body condition scoring of donkeys. Edinburgh, UK. Centre for Tropical Veterinary Medicine.

Pellegrini, P. 1999. On the Concept of Animal Breeds and Their Evolution in the Livestock Industry. Association of French Ruralists. *Ruralia.*

Piccione, G., Fazio, F., Caola, G., Refinetti, R., 2008. Daily rhythmicity in nutrient content of donkey milk. *Livestock Science,* **116**: 323-327.

Poivey, J. P., Landais, E., Seitz, J. L. 1980. Use of barymetry in local bull breeds of Côte d'Ivoire. *Journal of Animal Husbandry and Veterinary Medicine in Tropical Countries*, **33** (3): 311-317.

Polidori, P., Cavallucci, C., Beghelli, D., Vincenzetti, S. 2009. Physical and chemical characteristics of donkey meat from the Martina Franca breed. *Meat Science*. **82**: 469-471.

Quittet, E. 1976. The Goat: A Breeder's Guide. Paris: Editions la maison Rustique.

Raveneau, A., Daveze, J. 1996. The Donkey Book: Its History, Family, Training, and Life. Paris, France: Editions Rustica.

Rezende, M. P. G., Ramires, G. G., Souza, J. C. 2013. Are horses used for pulling carts in Aquidauana (MS) suitable for this purpose? *Dourados,* **6**, (22): 505-513.

RIM 1992. Nigerian National Livestock Resource Survey. Report by Resource Inventory and Management Limited (RIM). Abuja, Nigeria: FDL&PCS eds.

Roamba, R. C. 2014. Morphobiometric and biochemical characterization of donkeys (*Equus asinus*) in Senegal. Thesis submitted in partial fulfillment of the requirements for the degree of Doctor of Veterinary Medicine. Inter-State School of Veterinary Sciences and Medicine (EISMV). Cheikh Anta Diop University of Dakar. Faculty of Medicine, Pharmacy, and Dentistry.

Rossel, S., Marshall, F., Joris P., Pilgram, T., Adams, M., O'Connor, D. 2008. Domestication of the donkey: Timing, processes, and indicators. *Proceedings of the National Academy of Sciences,* **105** (10): 3715-3720.

Roux, M. 2006. Classification Algorithms. Paul Cézanne University, Marseille, France.

Salako, A. E. 2006. Application of morphological indices in the assessment of type and function in sheep. *International Journal of Morphology*, **24** (1): 13-18.

Salifou, K. 2014. Morphobiometric and biochemical characterization of donkeys (*Equus asinus*) in Burkina Faso. Thesis submitted in partial fulfillment of the requirements for the degree of Doctor of Veterinary Medicine. Inter-State School of Veterinary Sciences and Medicine (EISMV). Cheikh Anta Diop University of Dakar. Faculty of Medicine, Pharmacy, and Dentistry.

Salimei, E., Fantuz, F., Coppola, R., Chiofalo, B., Polidori, P., Varisco, G. 2004. Composition and characteristics of donkey milk. *Animal Research*, **5**: 67-78.

Sausseau, L. 1925. The Donkey, Mule-Breed Horses, and the Poitou Mule. Paris: Librairie des sciences agricoles.

Schaafsma, G. 2003. Nutritional significance of lactose and lactose derivatives: In H. Roginski, J. W. Fuquay, and P. F. Fox (Eds.). *Encyclopedia of Dairy Science*. **3:** 1529-1533.

Siméon, V. 2008. Donkeys & Mules - Introduction and Techniques for Care and Training, De Vecchi, France: Agricole Éd.

Smith, D. G., Pearson, R. A. 2005. A review of the factors affecting the survival of donkeys in semi-arid regions of Sub-Saharan Africa. *Tropical Animal Health and Production.* **37** (1): 1-19.

Stansfield, D. P. 1975. Genetics: Lectures and Problems. Paris, France: 1st ed. Schaum's Series, McGraw-Hill.

Starkey, P. 1994. Donkey utilization in sub-Saharan Africa: recent changes and apparent needs. Bakkoury M and Prentis R A (eds). Working Equines. Proceedings of the second international colloquium held April 20–22, 1994, Rabat, Morocco. Actes Éditions. pp. 289–302.

Starkey, P. 1997. Improving donkey utilization and management. Debre Zeit, Ethiopia. Report of the international ATNESA workshop held.

Starkey, P., Koorts, J. 1994. Animal traction in South Africa: the way forward. Debre Zeit, Ethiopia. Animal Traction Network for Eastern and Southern Africa: Report of the international.

Tavernier, R., Lizeau, C. 2000. Life and Earth Sciences. Paris: Bordas eds

Tesse, R., Claudia, P., Serena, B., Lucio, A. 2009. Adequacy and tolerance to donkey milk in an Italian cohort of children with cow's milk allergy. *Italian Journal of Pediatrics.* **35** (19): 1-4.

Tiringo, T. 2010. Investigation on nutritional and microbiological properties of Abyssinian donkey milk from Adea Woreda. Master of Science thesis in Food Science and Nutrition. Addis Ababa , University. School of Graduate Studies. Faculty of Science. Food Science and Nutrition Program.

Traore, A., Tamboura, H., Kabore, A., Yameogo, N., Bayala, B., Zare, I. 2006. Morphological characterization of small ruminants (sheep and goats) of the local Mossi breed in Burkina Faso.

Animal Genetic Resources Information. **39**: 3950.

Vall, E. 1996. Work capacity, exercise behavior, and physiological responses of zebu, donkeys, and horses in Northern Cameroon. Doctoral thesis., Ensam, Montpellier, France.

Vall, E. 1998. Work capacity of zebu, donkeys, and horses in northern Cameroon. Concept of animal-tool pairing suitability. *Annales de Zootechnie*, **47:** 41-58.

Vall, E., Ebangi, A. L., Abakar, O. 1998. Donkey-drawn farming: health and nutrition. Final report. Garoua, Cameroon. IRAD.

Vall, E., Ebangi, A. L., Abakar, O. 2001. Development of a body condition scoring system for draft donkeys in northern Cameroon. *Journal of Animal Husbandry and Veterinary Medicine in Tropical Countries*, **54**: 255-262.

▪▪▪

Verrier, E., Rognon, X., Laloë, D., De Rochambeau, H. 2005. Genetic tools and methods for the characterization, monitoring, and management of genetic variability in animal populations. *Ethnozootechnie*. **76**: 67-82.

Vigne, J. P. 1988. The major stages of goat domestication: A proposed explanation of its status in Western Europe. *Ethnozootechnie*. **41**: 1-13.

Vincenzetti, S., Polidori, P., Mariani, P., Cammertoni, N., Fantuz, F., Vita, A. 2008. Characterization of donkey milk protein fractions. *Food Chemistry*. **106**: 640–649.

Wilson, R. T. 1981. Distribution and importance of the domestic donkey in sub-Saharan Africa. Singapore. *Journal of Tropical Geography,* **2** (2): 136-143.

Wilson, T. R. 1990**.** The donkey. In W. J. A. Payne (Ed.), An introduction to animal husbandry in the tropics. Oxford: Blackwell Science Ltd.

Wold, A. G., Tegegne, A., Yami, A. 2004. Research needs of donkey utilization in Ethiopia. Donkeys, people and development: A resource book for the Animal Traction Network for Eastern and Southern Africa. Wageningen, Netherlands. Fielding & Starkey eds.

Yarkin, I. 1962. Atcilik, Ankara University Faculty of Agriculture Publications, No. 40, Ankara University Press, Ankara.

Yilmaz, O., Wilson, R. T. 2013. The domestic livestock resources of Turkey: notes on donkeys. *Journal of Animal and Plant Sciences*. **23** (2): 651-656.

APPENDICES

Appendix 1: Photos of some breeds from around the world

Poitou Donkey (France)

Catalan Donkey (Spain)

Mammoth Jackstock (United States)

Andalusian Donkey (Spain)

Pyrenean Donkey (France)

Miranda Donkey (Portugal)

Appendix 2: Results of the analysis of variance

Effect of location on body measurements

Source of variation	Dependent variables	Degrees of freedom	Mean Mean
Locations	Live weight	6	, , , „ " „ Ψ 666.010
	Body length	6	49.018
	Height at the withers	6	T . ^-t- 28.641
	Chest circumference	6	ψ 60,972
	Neck length	6	" ≈r- 67.636
	Neck circumference	6	⅜⅜ 127.837
	Head length	6	33.724**
	Head width	6	7.333**
	Ear length	6	6.380
	Chest depth	6	85.134**
	Chest width	6	31.777**
	Rump width	6	3.756
	Rump length	6	5.019
	Height at the croup	6	15.663
	Back height	6	26.774
	Barrel circumference	6	23.580
Residual error	Live weight	167	274.787
	Body length	167	36.281
	Height at the withers	167	12.308
	Chest circumference	167	24.531
	Neck length	167	9.502
	Neck circumference	167	15.409
	Head length	167	5.955
	Head width	167	1.650
	Ear length	167	3.135
	Chest depth	167	13.303
	Chest width	167	5.973
	Rump width	167	4,438
	Rump length	167	3.098
	Height at the croup	167	15.315
	Back height	167	14.043
	Barrel circumference	167	2.686
Total	Live weight	174	
	Body length	174	
	Height at the withers	174	
	Chest circumference	174	
	Neck length	174	
	Neck circumference	174	
	Head length	174	
	Head width	174	
	Ear length	174	

	Chest depth	174	
	Chest width	174	
	Rump width	174	
	Rump length	174	
	Height at the croup	174	
	Height at the withers	174	
	Cannon circumference	174	

: P < 0.05; : P < 0.01.

• **Effect of locality on biometric indices**

Source of variation	Dependent variables	Degrees of freedom	Sums of squares Mean
Location	Body mass index	6	0.004
	Body index (side view)	6	**T 7** * * 0.006
	Relative body index	6	^ **7** ** 0.010
	Compactness index	6	**. 7** * 0.049
	Dactylothoracic index	6	⅜⅜ 0.002
	Front-to-back height	6	0.001
	Cephalic index	6	**7 7** * * 0.006
	Pelvic index	6	0.001
Residual error	Body mass index	167	0.001
	Body index (profile)	167	0.002
	Relative body index	167	0.002
	Compactness index	167	0.020
	Dactylothoracic index	167	0.000
	Front-to-back height	167	0.000
	Cephalic index	167	0.001
	Pelvic index	167	0.002
Total	Body index	174	
	Body index (side view)	174	
	Relative body index	174	
	Compactness index	174	
	Dactylothoracic index	174	
	Front-to-back height	174	
	Cephalic index	174	
	Pelvic index	174	

: P < 0.05; : P < 0.01.

Appendix 3: Data collection form.

Location

Site

Origin Date..

Breeder's name ....

Gender Age

Qualitative variables

No. Order	1	2	3	4	5	6	7	8	9	10	11	12
Gender												
Color												
Age												
Presence of a cross												
Eye and nose color												
Stripes												

Height at the withers												
Body length												
Chest circumference Chest circumference												
Cannon circumference												
Neck circumference												
Height at the withers												
Head width												
Head length												
Neck length												
Chest depth												
Chest width												
Height at the withers												
Ear length												
Width at the croup												
Rump length												

Appendix 4: Dental chart for donkeys

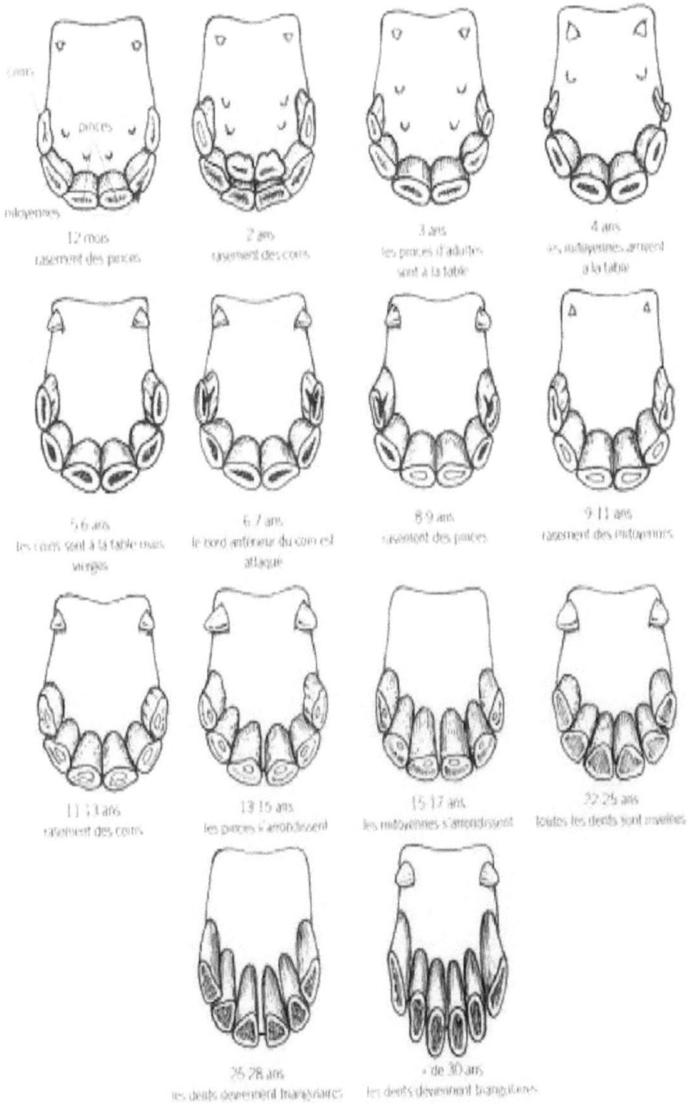

Appendix 5: Map of agroecological zones in Cameroon

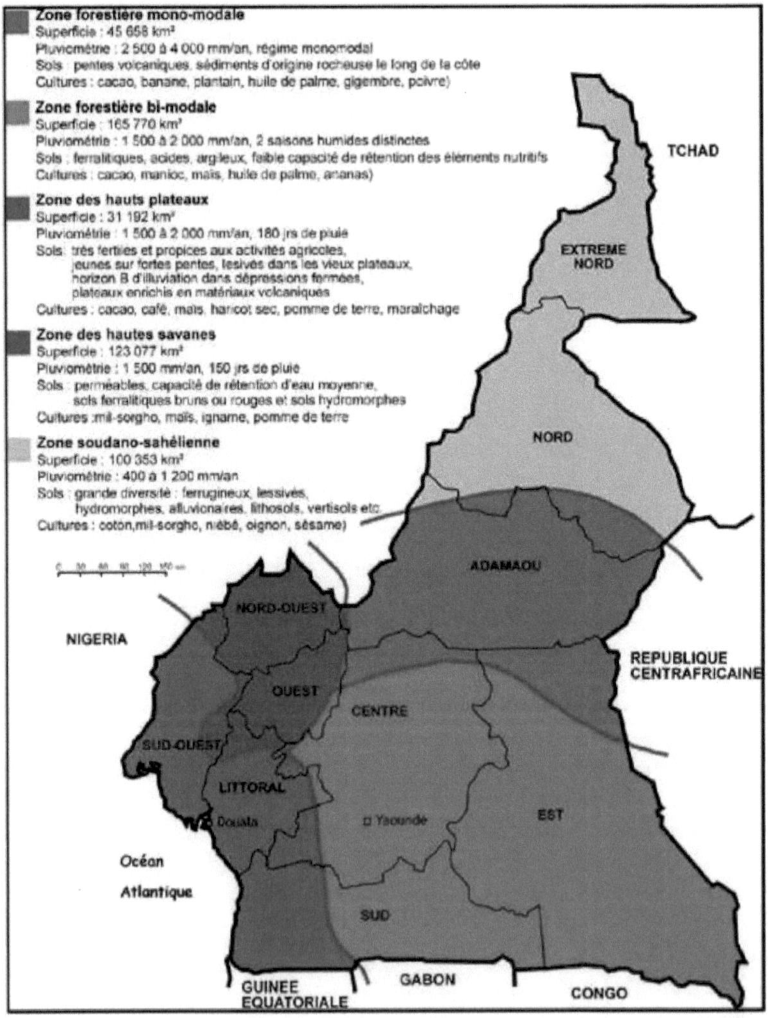

MIX
Papier aus verantwortungsvollen Quellen
Paper from responsible sources
FSC® C105338

Printed by Books on Demand GmbH, Norderstedt / Germany